Botany

A LAB MANUAL

Stacy Pfluger
Angelina College

JONES & BARTLETT
LEARNING

World Headquarters
Jones & Bartlett Learning
5 Wall Street
Burlington, MA 01803
978-443-5000
info@jblearning.com
www.jblearning.com

Jones & Bartlett Learning books and products are available through most bookstores and online booksellers. To contact Jones & Bartlett Learning directly, call 800-832-0034, fax 978-443-8000, or visit our website, www.jblearning.com.

Botany: A Lab Manual is an independent publication and has not been authorized, sponsored, or otherwise approved by the owners of the trademarks or service marks referenced in this product.

Production Credits
Chief Executive Officer: Ty Field
President: James Homer
SVP, Editor-in-Chief: Michael Johnson
SVP, Chief Marketing Officer: Alison M. Pendergast
Executive Publisher: Kevin Sullivan
Senior Acquisitions Editor: Erin O'Connor
Editorial Assistant: Rachel Isaacs
Editorial Assistant: Michelle Bradbury
Production Manager: Louis C. Bruno, Jr.
Senior Marketing Manager: Andrea DeFronzo
Production Services Manager: Colleen Lamy
Online Products Manager: Dawn Mahon Priest
VP, Manufacturing and Inventory Control: Therese Connell
Composition: Laserwords Private Limited, Chennai, India
Rights & Photo Research Associate: Lauren Miller
Title Page Image: © Cristi Matei/ShutterStock, Inc.

ISBN: 978-1-284-04106-4

6048

17 16 15 14 13   10 9 8 7 6 5 4 3 2 1

Brief Contents

Contents

Contents

Contents

Contents

Preface

Botany: A Lab Manual was written to provide the introductory botany student with an overview of plant biology. Plants are highly integrated, complex organisms that posses many structural adaptations that enable them to survive in many diverse habitats. Studying plants introduces you to an intriguing world of specialization and adaptation that has produced the wondrous variety of plant forms seen throughout the world. Though often overlooked and oversimplified, plants provide examples of fundamental metabolic processes, specialized environmental adaptations, and complex reproductive strategies.

Through my experience teaching undergraduate botany courses, I have come to appreciate the hurdles that many students face when beginning their journey through a typical botany course. The terminology can be cumbersome, and the concepts can be abstract. My goal with this manual was to explain the intricacies and complexities of plant anatomy and physiology in such a way that the new words and concepts not be stumbling blocks to comprehension. In doing so, I hope I have been able to simplify some of the confusing topics without sacrificing the accuracy necessary to a foundational biology course.

Critical thinking skills are essential to succeeding in scientific studies. To stimulate the development of such skills, critical thinking questions are embedded throughout the lab activities. Each lab is followed by brief guidelines for further study and concludes with a set of questions meant to encourage students to review what they have learned and to expand on that knowledge. I sincerely hope that this lab manual enhances, for all who use it, an appreciation for the plant kingdom, its diversity, and its contribution to life.

I would like to thank my students for providing the inspiration to take on this project and my husband for his continual support as I managed the stress of writing, meeting deadlines, and—my favorite of all—proofreading!

Stacy Pfluger
Angelina College

Chapter 1

Introduction to Botany and Microscopy

Laboratory Activities

Activity 1.1: Compound Light Microscope

Activity 1.2: Storing the Microscope

Activity 1.3: Stereomicroscope

Goals

Following this exercise students should be able to

- Identify the parts of a microscope.

- Understand the differences between the types of microscopes described.

- Use the microscope effectively.

Introduction

Botany is the scientific study of plants. It can be subdivided and incorporated into many fields: plant anatomy, plant physiology, ecology, cell biology, molecular biology, genetics, and many others. The study of plants can be approached as an experimental science using **controlled experimentation** to test **hypotheses**. A hypothesis is a plausible explanation of natural, observable phenomena that is testable. Human interest in plants began as a practical interest in obtaining more foods, fibers, and other plant-based goods for human use. Over time, curiosity about how plants work developed and gave life to the field of botany.

In many ways the advancements in botanical science are directly correlated to the available technology. Many tools are used in modern day botany; however, the microscope has had a large impact on all aspects of biology. Cells were first identified and named through the study of cork. Our understanding of the cell and its components increased as new technologies were developed.

Early microscopes were simple designs that relied on a single lens to magnify the object. These microscopes provided limited magnification, approximately equal to what one might experience with a hand lens. The **compound light microscope** is one of the most commonly used microscopes. A compound microscope sends a beam of light through a thin section of a specimen and uses multiple lenses to enlarge the image. A series of lenses is beneficial because the second lens compounds the magnification of the first. Most compound light microscopes are used to magnify images up to 1,000×.

Much magnification beyond this limit causes a problem with **resolution**. **Magnification** simply refers to the process of making an object appear larger. Resolution refers to the ability of a lens to distinguish between two closely adjacent points. Beyond a certain magnification, increased magnification does not result in increased resolution. Therefore, even though the image would increase in size, objects would not become any clearer.

Although widely used, compound light microscopes are limited in their effectiveness. They rely on light passing through a thin section of the specimen. These microscopes are not as useful for observing the surface of intact, multicellular organisms. For those applications a **stereomicroscope**, also called a dissecting microscope, is useful. Typical stereomicroscopes can magnify images from about 5× to 40×. Though the magnification is much lower than that of a compound light microscope, stereomicroscopes are useful. Stereomicroscopes are almost always binocular, having two eyepieces through which one observes the specimen. This enhances the three-dimensional appearance of the specimen. Conversely, compound light microscopes provide a two-dimensional view of an object. Student compound microscopes may be monocular or binocular.

When higher magnification with high resolution is needed, light microscopes are insufficient. Instead, electron microscopes can be used. Electron microscopes use a beam of electrons to form the image being magnified. Electrons allow for much higher resolution, and electron microscopes provide highly magnified images with exceptional resolution.

A **transmission electron microscope** is analogous to a compound light microscope. It provides detailed images of the internal structure of a specimen. The beam of electrons must pass through the specimen, so each specimen must be carefully prepared and sliced with a diamond knife to generate extremely thin sections.

The **scanning electron microscope** provides detailed image of the surface of a three-dimensional object. Electrons bounce off the surface of the specimen instead of passing through it; therefore, the specimen does not need to be sliced and may be left intact. The primary advantage of a scanning electron microscope is the generation of a highly detailed, clear image of a structures surface.

Although both types of electron microscopes have provided a wealth of information about cell structure, many scientists still rely on the less expensive and more common light microscope. Throughout this semester you will be using microscopes extensively to study plant anatomy and physiology. Typically, you will be using compound microscopes; however, stereomicroscopes can be used to observe some of the larger structures. In this lab we examine both types of light microscopes and their uses.

Activities

■ Activity 1.1: Compound Light Microscope

The compound light microscope is a familiar fixture in many biology labs. Many are binocular containing two **eyepieces**, whereas others are monocular and have only one eyepiece. As you look into the eyepiece, you will see the **ocular lens**. The ocular lens typically has a magnification of 10×. This should be labeled on the outside of the eyepiece. Below the eyepiece you will find the **body**, which is the housing around the internal structure of the microscope. It connects the ocular lens to the objective lens and keeps the lenses aligned. The objective lenses are found just above the stage, where the specimen is placed. Each microscope has three or four **objective lenses** attached to a revolving nosepiece. The objective lens in use may be changed by twisting the revolving nosepiece until the next objective clicks into place. The objective lenses have specific names. The **scanning power objective** has the lowest magnification, usually 4×. It is also the shortest objective. The **low power objective** is somewhat longer and has a magnification of 10×. The **high power objective**, sometimes called high-dry power, has 40× magnification. A fourth objective, if present, is an **oil-immersion objective** with a magnification of 100×. This level of magnification is typically not needed to view botanical specimens. The **total magnification** of an image as it reaches your eye can be calculated by multiplying the magnification of the objective lens in use by the magnification of the ocular lens. For example, if you used the scanning lens (4×) to observe a specimen, total magnification is 40× (total magnification = 4× times 10× = 40×).

The part of the microscope that sits directly on the table top is the base. Within the base is the **illuminator**; in most modern microscopes this is a light bulb. The arm of the microscope extends upward from the base and supports the eyepieces. The stage is controlled by two focus adjustment knobs. Typically, these knobs are stacked on one another and are found on both sides of the arm, near the base. The **coarse adjustment knob** is the larger of the two and moves fairly rapidly in a vertical plane. It is used for focusing on an object with the scanning or low power objectives. The smaller of the two knobs is the **fine adjustment knob**. It moves the stage vertically as well; however, it moves the stage much more slowly. On the upper surface of the stage is a metal stage clip. The **stage clip** is composed of a fixed metal bracket on one side with a moveable lever (clip) on the other side. When used properly it holds the slide in place and makes the mechanical stage more useful. The stage is controlled mechanically by a set of stacked knobs, called the **mechanical stage controls,** found on the lower side of the stage. The upper knob moves the stage forward and backward, whereas the lower knob moves the slide from left to right.

Just beneath the stage is the **condenser**. It can be seen through the opening in the center of the stage. The condenser is a lens that focuses light from the illuminator onto the specimen on the slide. It is controlled by a **condenser control knob** found beneath the stage near the arm. In the same area with the condenser is the **iris diaphragm**. The iris diaphragm controls how much light reaches the specimen. Most microscopes have a small lever present on the side of the condenser that controls the diaphragm. Some microscopes have diaphragms that are controlled by turning the condenser housing. Your instructor will demonstrate use of the diaphragm on your microscope.

When carrying a microscope always use two hands. Place one hand on the base and one on the arm. It is important to be careful because these are expensive pieces of equipment.

1. Acquire your assigned compound microscope per your instructor's directions.
2. Identify all the parts of the microscope described above.
3. Provide the total magnification for each lens below (assuming the current ocular lens is in use).

   a. Scanning

   b. Low power

   c. High power

   d. Oil immersion (if present)

Each objective lens has a different **field of view**. The field of view is simply the size of the area that can be seen at any one time using a particular lens. The lenses in a microscope invert the image as it is magnified. This is due to the nature of light reflection by a lens.

4. Obtain a prepared slide of the letter "e."

5. Observe it under scanning power. Use the coarse adjustment knob to bring the specimen into focus. Always begin focusing on an object with the scanning objective. Diagram what you see below.

6. Compare the orientation of the letter "e" as seen with the microscope with the orientation seen with the unaided eye.

7. Switch to low power. Do not move the focus knobs or mechanical stage. Simply turn the revolving nosepiece until the low power objective clicks into place. If it does not click into place, you will not be able to see light through the eyepiece. You should be able to bring the specimen into sharp focus with just a few turns of the fine adjustment knob. This is because the microscope is **parfocal**. When an object is centered and in focus with one objective, it remains roughly centered and in focus when the objective is changed.

8. How does the letter "e" appear different under low power as opposed to scanning power?

9. Diagram the letter "e" as seen on low power.

10. Now switch to high power. How does the letter "e" appear now?

Notice, as you moved from relatively low magnification objectives up to the high power objective, the distance between the objective and the slide decreased. This is called the working distance. The **depth of field** is another important concept in microscopy. Depth of field refers to the thickness of the specimen that is in focus at any point in time.

11. Obtain a prepared slide of crossed threads. Move the slide so the crossed portions of the threads are in view.

12. Observe the specimen under scanning power. How many threads are in focus?

13. Switch to low power. How many treads are in focus now? Can you tell which is on top of the others?

14. Switch to high power. How many threads are in focus at one time? Which one is on top?

As you moved from the lowest magnification to the highest magnification, you should have noticed that fewer threads were in focus at a given point in time. Objectives that have a large depth of field allow you to see thick objects, or multiple threads, with a high percentage of the specimens in crisp focus at one time. Conversely, objectives with low depth of field allow you to see small portions of the specimen in focus at one time.

Observe any other specimens your instructor has out for display.

COMPARE AND CONTRAST

1. Which objective has the shortest working distance?

2. Which objective has the smallest depth of field?

3. Which objective has the largest field of view?

■ Activity 1.2: Storing the Microscope

Storage is an important component of caring for a microscope to ensure it is in working order the next time you use it. Follow the steps below to properly store your microscope.

1. Turn off your microscope and unplug it.
2. Lower the stage.
3. Wrap the cord around the base.
4. Place the scanning objective in the operating position.
5. Carefully carry it to the proper storage location.

■ Activity 1.3: Stereomicroscope

Stereomicroscopes are relatively simple in design. They have many of the same parts as a compound microscope. They are typically equipped with binocular eyepieces and objective lenses that are located on a turret. The objective turret twists to allow you to change objective lenses. The base has a stage plate and usually has one or two stage clips to anchor the slide. These microscopes may have two illuminators. Sometimes one is present in the base, beneath the stage plate. Most of them have an illuminator just below the objectives that shines light on the surface of a large object.

1. Pick up your stereomicroscope as directed by your instructor.
2. Identify the parts of the microscope.
3. Obtain a prepared slide of the letter "e." Observe it under the lowest magnification. Diagram what you see.

4. How does the letter "e" compare as viewed through the microscope versus how it appears to your unaided eye?

....

....

....

5. Switch to a higher magnification. How does the "e" appear now?

....

....

....

6. Observe any other specimens your instructor has on display.

COMPARE AND CONTRAST

1. How does the letter "e" appear differently with a stereomicroscope as opposed to a compound microscope?

....

....

....

....

2. Describe the benefits of the compound microscope and the stereomicroscope.

....

....

....

....

Study Guide

- Be able to define the terms in bold.
- Be able to describe the correct operation of the compound light microscope.
- Be able to differentiate between the lenses on the compound light microscope and their characteristics.

Conclusions

1. What magnification is provided by the scanning, low power, and high power objectives?

2. What is the field of view? How does it change with increasing magnification?

3. What is the depth of field? How does it change with increasing magnification?

4. What is the working distance of an objective? How does it change with increasing magnification?

5. What is the advantage of a parfocal microscope?

6. What is resolution? How does it impact the maximum available magnification of a microscope?

7. Describe the difference in how specimens appear with a stereomicroscope versus a compound microscope.

8. If you wanted to see a highly detailed image of a pollen grain, which type of microscope would be optimal? Why?

....

....

....

9. You are observing a *Paramecium* swimming in pond water using the low power objective on your lab microscope. What is the total magnification of the *Paramecium* as you see it?

....

....

....

10. Describe three differences between your lab's compound microscopes and stereomicroscopes.

....

....

....

Chapter 2

Plant Cells

Laboratory Activities

Activity 2.1: Cork Cells

Activity 2.2: Onion (*Allium*) Epidermis

Activity 2.3: *Elodea* Leaf

Activity 2.4: Potato Tuber Cells

Activity 2.5: Tomato Pulp Cells

Activity 2.6: Tomato Epidermis

Activity 2.7: *Zebrina* Stem Tissue

Activity 2.8: Osmosis and Plant Cells

Goals

Following this exercise students should be able to

- Identify the major parts of a plant cell.
- Describe the functions of the major plant organelles.
- Make a wet mount of a sample.
- Understand osmosis and its importance in plants.
- Distinguish between hypotonic, hypertonic, and isotonic solutions.

Introduction

In this exercise you will learn about the structure of plant cells and the function of the various organelles. Plants are eukaryotes, meaning the cells have a membrane-enclosed **nucleus** as well as many other intracellular compartments called **organelles**. Each organelle functions to help compartmentalize the cell's activities, allowing it to become more efficient.

For this exercise you need to make **wet mounts** of your samples. For each item obtain a small piece of tissue (the smaller, the better). Place the sample on the surface of a clean, dry slide. Add one to two drops of water (or dye) to the sample. Carefully place a coverslip on top of the sample and water. To avoid excess air bubbles under the coverslip, hold the coverslip at a 45-degree angle above the slide surface. Lower the coverslip until it touches the edge of the liquid and drop the coverslip. Your instructor will demonstrate this technique.

You will be observing the following samples: cork cells, onion epidermis, *Elodea* leaf, potato tuber cells, tomato pulp, tomato epidermis, and *Zebrina* stem tissue.

Activities

▪ Activity 2.1: Cork Cells

Cork cells make up the bulk of the outer bark of a woody plant. These cells are dead; however, they continue to function in supporting the plant and protecting it from pathogens and desiccation. A fully formed cork cell will accumulate large amounts of water-proofing compounds in their cell wall, which eventually leads to the cell's death. At maturity the cells are dead and have no cytoplasm or any internal structures.

The **cell wall** is the outermost part of a plant cell. It provides support to the cell and helps prevent osmotic lysis, as you will see later in this lab.

1. Cut a thin section of cork and make a wet mount. The thinner the section, the better you can see individual cells.
2. Observe the cells under low power. Locate an area of the specimen where the cells are one or a few layers and can easily be viewed in sharp focus.
3. In the space provided, draw a few cork cells. Label the cell wall.

▪ Activity 2.2: Onion (*Allium*) Epidermis

The onion bulb is composed of layers of thick leaves. Bulbs are modified as storage organs for the plant. The leaves are the storage compartment of the bulb. They surround a short, central stem. The layer you want to observe is the covering of the leaf, called the epidermis.

1. Obtain a small sample of onion epidermis, taken from the inside of one onion bulb scale. This tissue is thin and has a tendency to curl at the edges. Be sure to place it flat on the surface of the slide. Add one to two drops of IKI (iodine-potassium iodide) solution to the onion epidermis. The iodine provides increased contrast and makes the nucleus easier to see.
2. Examine the slide with low power. Find a region of the sample where the cells are flat on the surface of the slide and the area appears focused. Switch to high power.
3. Try to locate the following structures.
   a. Cell wall
   b. Cytoplasm
   c. Nucleus
   d. Nucleolus

The **cell wall** is the outermost part of a plant cell. It provides support to the cell and helps prevent osmotic lysis, as you will see later in this lab. Inside the cell wall is a plasma membrane, which is not visible at this magnification. The plasma membrane encloses the living part of the cell, the **cytoplasm**. The cytoplasm contains a fluid portion, called the cytosol, as well as numerous organelles.

The most prominent structure inside the cell is the **nucleus**, which contains the cell's genetic information. Within each nucleus you should see one or a few nucleoli. Each **nucleolus** is a region of the nucleus responsible for producing ribosomal RNA, which forms part of the ribosome (the organelle responsible for producing proteins). The nucleoli appear as dark spots within each nucleus.

These are storage cells, so each should have a large central vacuole. The vacuole is a membrane-bound structure that contains fluids and dissolved substances. These substances vary from cell to cell but are often referred to as the cell sap. The vacuole in these cells appears as a light-colored area in the center of the cell. The vacuole is surrounded by the **vacuolar membrane**, sometimes referred to as the tonoplast. You cannot see the vacuole directly, but you should be able to determine its location by the absence of any other structures in this area.

4. Diagram a single epidermal cell in the following space. Label the **cell wall**, **nucleus**, **nucleolus**, and **central vacuole**.

COMPARE AND CONTRAST

1. How does the shape of cork cells differ from that of epidermal cells?

2. How can both of these cells continue to function when one is alive and the other is dead?

3. Describe one similarity in the functions of these two cell types.

■ Activity 2.3: *Elodea* Leaf

Elodea is an aquatic plant that is widespread in pond environments. It is often called Anacharis. It lives entirely submerged under water and is a common addition to aquariums. The leaves of *Elodea* are quite thin at only two cell layers thick. This makes *Elodea* an ideal specimen for observing photosynthetic cells.

1. Make a wet mount of an entire *Elodea* leaf.
2. Examine it on low power. The best area to observe is between the edge of the leaf and the midrib (vein). Once in focus, switch to high power.
3. Identify the following structures:
   a. Cell wall

   b. Cytoplasm

   c. Central vacuole

   d. Chloroplasts

   e. Nucleus

Chloroplasts are one example of a **plastid**. Plastids are a group of organelles found in plant cells and some protists. These plastids are surrounded by multiple membranes, contain DNA, control their own replication cycles, and perform specialized functions within the cell. Throughout the next several samples you will be introduced to several types of plastids.

The **chloroplast** is the organelle that is responsible for photosynthesis. **Chlorophyll** is a green pigment that is primarily responsible for absorbing light during photosynthesis. Chloroplasts contain large quantities of chlorophyll and will typically appear green. They are ovoid organelles surrounded by two membranes. Inside the outer membranes are a third set of membranes forming stacks called **grana**. These are the **thylakoid membranes** that actually contain chlorophylls. Because each chloroplast contains such large numbers of grana, the entire organelle will appear green at this magnification.

In some cells you should be able to see the chloroplasts moving around the periphery. This phenomenon is due to **cytoplasmic streaming**, the constant flow of cytoplasm within a cell. Many substances are moving within the cytoplasm, but the chloroplasts are simply easier to observe than most.

The nucleus is very difficult to see in these cells. It is colorless and requires careful adjustment of the fine focus to be seen. The **central vacuole** is easy to identify by the lack of chloroplasts in the central region of the cell. The chloroplasts are near the cell wall because the central vacuole occupies most of the central cytoplasm within the cell.

4. Diagram a typical *Elodea* cell in the space below. Label the cell wall, cytoplasm, chloroplast, and central vacuole.

COMPARE AND CONTRAST

1. The previous two types of cells you have seen are from leaves. These cells appear to be quite different from one another. Why do you believe that may be? Hint: think about the functions of these cells.

2. Would you expect to find chloroplasts in the cells of the onion bulb scale directly (not the epidermis coving the scales). Why or why not?

■ Activity 2.4: Potato Tuber Cells

White potatoes are modified stems that grow underground and are used for storing starch as a food reserve for the plant. These modified stems are called tubers. Because these stems are modified for storing starch, they are composed primarily of parenchyma cells. As you observe your specimen, pay attention to the spacing and arrangement of the cells. In the *Elodea* leaves and onion epidermis, you saw tightly packed cells. In the potato tuber the cells are farther apart and less organized. This results in **intercellular spaces** between the cell walls of adjacent cells.

Inside these parenchyma cells are numerous plastids. Plastids that function in storing complex molecules, other than pigments, are termed **leucoplasts**. Typically, leucoplasts are numerous and appear as small ovoid structures within the cell. Those that specifically function in starch storage are **amyloplasts**.

1. Cut a small, thin section from the interior of a potato (not the "skin"). Make a wet mount. Note: if your sample is too thick, it will be difficult to see the individual cells and amyloplasts.

2. Examine the slide with low power. Find an area where you can see intact cells clearly. The best place to look is toward the edge of the specimen.

3. Look within the cells to see the amyloplasts. You may want to use high power to view the amyloplasts in more detail; however, they are numerous and can be difficult to focus on clearly with high power.

4. If you have difficulty seeing the amyloplasts, add a small drop of IKI to the specimen. Iodine turns starch dark purple. This will make the entire specimen quite dark, but you will be able to visualize how much starch is present within these cells.

5. Diagram a few potato tuber cells below. Label the cell walls, amyloplasts, and intercellular spaces.

COMPARE AND CONTRAST

1. How does the shape of the potato tuber cells differ from that of the onion epidermal cells? Explain how the shape and spacing of these cells is advantageous to the plant organs in which they are found.

....

....

....

....

....

2. Why is the central vacuole not as prominent in the potato tuber cells relative to the *Elodea* leaf cells?

....

....

....

....

....

■ Activity 2.5: Tomato Pulp Cells

Many fruits and flowers are brightly colored. These red and yellow colors result from the presence of specific pigments within the cells of these organs. Those pigments are contained within **chromoplasts**, plastids that contain pigments other than chlorophyll. Chromoplasts vary in their intracellular location; however, there is usually a cluster that forms around the nucleus.

1. Make a wet mount of the pulp of the tomato fruit. The region closest to the epidermis (the "skin") is best. Do not include any of the tomato epidermis in your sample.
2. Observe the cells under low power. Once you have focused on an area of the sample where the cells can be clearly seen, switch to high power.
3. Draw a tomato pulp cell. Label the cell wall and chromoplasts.

COMPARE AND CONTRAST

1. Do these cells more closely resemble those of the potato or of the onion?

2. Compare the size, shape, and color of chromoplasts with chloroplasts and amyloplasts.

■ Activity 2.6: Tomato Epidermis

The epidermis of the tomato is the outer covering of the fruit and is often referred to as the "skin." These cells, and many others in plants, are interconnected by small extensions of the plasma membrane that extend through holes in the cell wall, called plasmodesmata. In some plant cells plasmodesmata are clustered in a region called a **primary pit field**. The primary pit field is an area of the cell wall where the cell wall is particularly thin.

1. Make a wet mount of a thin section of tomato epidermis.
2. Observe the cells under high power. It is best to locate the thinnest region of the specimen, usually toward the torn edge.
3. In the space below, draw two tomato epidermal cells. Label the cell wall, cytoplasm, and plasmodesmata.

COMPARE AND CONTRAST

1. Which cells are more similar in appearance to the tomato epidermal cells, the onion or the potato? Why do you believe this is so?

2. How do the cell walls of tomato pulp cells differ from those of the tomato epidermal cells?

■ Activity 2.7: *Zebrina* Stem Tissue

Plants do not have an excretory system to carry away metabolic wastes. Instead, many plants avoid harmful accumulation of these toxic metabolic wastes by converting them to insoluble substances. These insoluble forms crystallize and can be stored in the plant cell without damaging the cell. One of these metabolic wastes is oxalic acid, which can be converted to calcium oxalate crystals.

Some crystals function as defensive mechanisms for plants. When these plant tissues are chewed by humans or other animals, the crystals of calcium oxalate and other compounds produce a burning sensation and can lead to inflammation and swelling of the mouth and airways.

Zebrina is a common plant used for groundcovers and hanging baskets. Some varieties are variegated, some have purple leaves and stems, and some are low-growing green plants. *Zebrina* produces long, thin, needle-like crystals of calcium oxalate. These crystals are usually referred to as **raphides** because of their elongated shape. These crystals cannot be seen within intact cells using a compound light microscope; however, they can be seen extending out of damaged cells.

1. Make a wet mount of a thin cross-section of *Zebrina* stem.
2. Observe the slide under low power. The raphides are most visible along the cut edges of the stem. They are colorless and appear as thin, elongated structures, usually with pointed ends.
3. If you cannot observe the raphides, use a razor blade to chop the stem section on the slide. Add a drop of water, replace the coverslip, and observe under low power.
4. Sketch a few raphides below.

COMPARE AND CONTRAST

1. How are raphides different from plastids? Refer to their respective functions and structures.

2. Why would you not expect to find raphides in tomatoes or peppers?

■ Activity 2.8: Osmosis and Plant Cells

Osmosis is the diffusion of water across a selectively permeable membrane. The plasma membrane of cells is a selectively permeable membrane that allows free passage of some substances and limits the passage of others. In this exercise you will observe the effects of osmosis on plant cells.

The relative concentrations of solutes in two areas can be compared using **tonicity**. In biology we typically use tonicity to compare solute concentrations inside the cell with those of the cell's surroundings. An isotonic solution has the same concentration of solutes as is found within the cell's cytoplasm. When a cell is placed in an **isotonic** solution, water molecules move across the cell membrane in both directions at the same rate. The cell experiences neither a net loss nor a net gain of water. A **hypertonic** solution contains more dissolved solutes than does the cytoplasm. A cell placed in a hypertonic solution experiences **plasmolysis** due to a net loss of water as water osmoses from the cytoplasm into the solution surrounding the cell. **Hypotonic** solutions contain a lower solute concentration than is present within the cytoplasm. A cell placed in a hypotonic solution becomes **turgid** because of a net gain of water as water osmoses from the solution into the cytoplasm. As water enters the cell, osmotic pressure, or **turgor pressure**, can build up within the cell. Water will continue to enter the cell until this turgor pressure pushing outward on the cell wall equals the resistance of the cell wall pushing back on the cell membrane and cytoplasm. This situation is characteristic of a turgid cell.

1. Obtain another *Elodea* leaf.
2. Make a wet mount of this leaf with distilled water. Distilled water is hypotonic to the cytoplasm.
3. Sketch a representative cell below.

4. Now add 10% NaCl (sodium chloride) solution to the slide. This can be accomplished by applying the NaCl solution to the edge of the coverslip. The salt solution will flow under the coverslip due to capillary action. A 10% NaCl solution is hypertonic to the cytoplasm.
5. Sketch a representative cell below.

6. How can you explain the differences in the cells drawn in numbers 3 and 5 above?

Study Guide

- Be able to define the terms in bold.
- Be able to identify the structures labeled in the drawings.
- Be able to describe the function and location of the cell structures observed during this lab.
- Be able to answer each of the questions asked in the lab exercise.

Conclusions

1. What is the primary role of the cell wall? How can you use this information to explain the differences in the thicknesses of the cell walls seen in the various specimens?

2. What is a plastid? What characteristics are shared by all the plastids studied today?

3. What is an epidermis? How does the epidermis differ from cells deeper within a plant organ like a fruit?

4. Describe the role of plasmodesmata? How do plasmodesmata impact the individuality of the cells?

5. Compare and contrast chromoplasts and chloroplasts with respect to cell type, color, and function.

6. Is an amyloplast more similar to a chloroplast or a chromoplast? Why?

7. Describe two benefits of crystal formation in plants.

8. Describe the impact of hypotonic, isotonic, and hypertonic solutions on plant cells. Which is most likely to lead to the death of a plant cell after prolonged exposure?

Chapter 3

Cell Division

Laboratory Activities

Activity 3.1: Mock Mitosis

Activity 3.2: Mitosis in Onion Cells

Activity 3.3: Mock Meiosis

Goals

Following this exercise students should be able to

- Recognize the stages of mitosis.
- Understand the importance of mitosis in the cell cycle.
- Explain the role of mitosis in growth and repair of the plant.
- Differentiate between cytokinesis and karyokinesis.
- Describe the key events of each stage of mitosis.

Introduction

All cells have a life cycle, which we call the **cell cycle**. The cell cycle begins with the formation of a new cell and continues until that cell goes through cell division to produce its own daughter cells. **Cell division** is a process by which one cell (mother cell) divides to produce two new cells (daughter cells). It is composed of two primary phases: **karyokinesis** and **cytokinesis**. Karyokinesis refers to the division of the nucleus, whereas cytokinesis refers to the division of the cytoplasm.

   There are two types of karyokinesis. The most common form, **mitosis**, produces daughter cells that are identical to the mother cell. They have the same number of chromosomes and the same genetic content. Mitosis is critically important for the growth and repair processes of a plant. The second type is **meiosis**, which results in the production of four genetically different daughter cells each of which contain half the number of chromosomes present in the mother cell. Meiosis is typically considered to be a part of the reproductive cycle of an organism. By reducing the number of chromosomes by half, the cells produced are **haploid** (having only one copy of each chromosome). When these cells fuse together, during fertilization (or syngamy), they form a **zygote** that is **diploid** (having two copies of each chromosome). The use of meiosis to produce a haploid cell, therefore, ensures that with each successful reproductive attempt the chromosome number stays the same. In other words, without meiosis, if a cell with four chromosomes fertilized another cell with four chromosomes, the resulting zygote would have eight chromosomes. Repeat this process again, and the chromosome number increases to 16. The total number of chromosomes would increase each time an organism reproduced sexually. Meiosis prevents this and helps to ensure the chromosome number remains constant in the species.

The division of the nucleus is much more tightly regulated than the division of the cytoplasm. Mitosis is typically divided into four stages, each of which is marked by chromosomal events: prophase, metaphase, anaphase, and telophase. Before mitosis can begin, the genetic content of the cell must be duplicated so each daughter cell can inherit an exact copy of the mother cell's DNA. This occurs during **interphase**, defined as the time in the cell cycle between divisions. Most plant cells spend most of their life cycle in interphase. It is a period of growth and preparation for division. Once the DNA is replicated during interphase, the cell then produces microtubules and other substances necessary for mitosis.

The first stage of mitosis is **prophase**. During prophase the nuclear envelope begins to disintegrate, the nucleoli disappear, and the **chromosomes** condense. There is no specific pattern to the chromosomes during prophase. **Spindle fibers** (microtubules) attach to each chromosome beginning in prophase. Because the DNA of the cell has already doubled through DNA replication in interphase, the chromosomes have a distinct structure. At this point each chromosome contains two identical **sister chromatids**. The chromatids are attached to one another by a **centromere**. It is this centromere that attaches to the spindle fibers.

Prophase is followed by **metaphase**. By now, the spindle fibers have all attached to chromosomes and have pulled the chromosomes to the center of the cell where they align at the cell equator, sometimes called the metaphase plate. The spindle fibers are attached to each centromere, so the centromere is what aligns in the center of the cell. The chromatids will not align perfectly in the center of the cell.

Once the chromosomes align in metaphase, the centromeres begin to break down, freeing the sister chromatids. The spindle fibers shorten, pulling the newly freed sister chromatids, now called **daughter chromosomes**, toward opposite poles of the cell. This is **anaphase**. When the chromosomes reach the poles of the cell, a new nuclear envelope begins to reform, nucleoli reappear, and two new nuclei are formed. This occurs during **telophase**. Telophase is usually accompanied by cytokinesis, which results in two new daughter cells, each with a diploid nucleus identical to that of the original mother cell. There are instances when cytokinesis does not accompany telophase. This results in a multinucleate cell. Keep in mind this is a continual process. It is easier to study the overall process by looking at each stage separately; however, the cell moves continually from one stage to the next.

In plant cells cytokinesis includes the formation of the new cell wall that will border the daughter cells. The **cell plate** forms in the center of the cell and forms the basis of the new primary cell walls. Because this usually occurs during telophase, observing the cell plate can often be useful as a marker for telophase. The contents of the cytoplasm are divided randomly between the two new daughter cells. Organelles are divided into both cells, but each cell may acquire different numbers of chloroplasts or mitochondria. As long as each daughter cell has at least one of each organelle, both can function and produce more organelles as needed.

Remember, chromosomes are only visible during cell division when they are condensed. When the cell is not actively dividing, chromosomes are found dispersed throughout the nucleus in the form of **chromatin**. Chromosomes are composed of DNA, the genetic material of the cell, and histone proteins around which the DNA winds.

Activities

▪ Activity 3.1: Mock Mitosis

Your first priority in this lab is to learn the stages of mitosis. Once you understand what happens in each, especially with regard to the chromosomes, you'll be better equipped to identify cells in each stage.

1. Obtain a piece of scratch paper, eight strips of colored paper (four each of two colors) representing chromosomes, one piece of string approximately 12 inches long, and four paper clips from your instructor.

2. The scratch paper is your cell. Form a circle with the string inside the "cell" to represent the nuclear envelope.

3. Place two strips of paper (one each of two colors) into the nucleus. This represents the cell during early stages of interphase.

4. Mimic DNA replication. The cell must make an additional copy of its DNA. To do this, attach a second strip of paper (same color) to each of the existing "chromosomes" using the paper clips. You now have four chromosomes, each composed of two chromatids.

5. Move your cell's components to mimic prophase. What object did you remove?

6. Move the components to the appropriate conditions for metaphase. Remember, the centromeres align at the cell equator.

7. To mimic anaphase, remove the paper clips that hold your sister chromatids together. That symbolizes the centromere breaking down. Now move the new daughter chromosomes toward the poles of the cell.

8. To mimic telophase, fold the string in half to form two new nuclear envelopes.

9. Cytokinesis usually occurs at the same time as telophase. Draw the cell plate in the middle of the paper.

COMPARE AND CONTRAST

1. How are the daughter cells similar to one another? How are they different?

....

....

....

....

2. How are the daughter cells similar to the original mother cell? How do they differ?

....

....

....

....

▪ Activity 3.2: Mitosis in Onion Cells

You will be observing the stages of mitosis in onion root tip cells. The actively reproducing cells in an onion root can be found just inside the tip of the root and extending up either side of the root near the outer surface. The cells have been stained with a dye that makes the DNA appear dark, usually pink or purple in color. By observing the locations of the chromosomes, you should be able to identify cells in each stage of mitosis.

1. Obtain a prepared slide of onion (*Allium*) root tips.
2. Observe the cells under low power.
3. Locate a cell in each of the stages of mitosis: prophase, metaphase, anaphase, and telophase. Sketch an example of each below.

4. Locate a cell that is not currently dividing. This cell will be in interphase.
5. Sketch an example of a cell in interphase.

COMPARE AND CONTRAST

1. How do the cells you observed compare with the one you generated in Activity 3.1?

....

....

....

....

2. Compare and contrast the appearance of a plant cell in interphase with one in prophase.

3. What does a cell in telophase have in common with a cell in interphase?

4. In looking at your slide of the onion root tip, most of the cells were in one phase of the cell cycle. What is that phase? Why is that not surprising?

▧ Activity 3.3: Mock Meiosis

Meiosis has the same four main stages as seen in mitosis: prophase, metaphase, anaphase, and telophase. The difference is that meiosis involves two successive divisions of the nucleus. During the first division, **meiosis I**, the number of chromosomes is reduced; thus, it is sometimes referred to as a reduction division. During the second division, **meiosis II**, the sister chromatids formed during interphase are separated; however, chromosome number remains the same. It is therefore called an equational division. Because there are two divisions, each stage (prophase through telophase) occurs twice. Roman numerals differentiate between the two divisions. Prophase I, for example, occurs during meiosis I, whereas prophase II occurs during meiosis II.

1. Find one of your daughter cells from the mock mitosis exercise. Cut along the cell plate line you drew earlier to separate the cells. Remove the nucleus from one cell and cut that piece of paper (cell) in half again (for a total of two quarter-page pieces and one half-page piece of paper with intact nucleus).
2. The intact cell will now be going through meiosis.
3. Mimic DNA replication as before. Use paper clips to attach the sister chromatids to one another.

4. Mimic prophase I by removing the nuclear envelope. During prophase I homologous chromosomes pair up in a process called **synapsis**. To represent synapsis, pair up the two chromosomes of matching colors. These pairs are called **tetrads**.

5. **Crossing-over** typically occurs in which nonhomologous members of a tetrad exchange portions of chromatids. On the chromatids of one chromosome, write the letter "A" at the top edge. On the chromatids of the other chromosome, write the letter "a" at the top edge. Now, cut the ends off of one chromatid from each chromosome and switch places with them. Tape them in place. You have now mimicked crossing-over.

6. Move the chromosomes to their positions at the cell equator. In metaphase I the tetrads align at the cell equator with one chromosome on either side of the equatorial line.

7. During anaphase I the homologous chromosomes separate, so move one chromosome from each tetrad toward opposite poles of the cell. Notice that the centromere is still intact and each chromosome is composed of two sister chromatids.

8. Move the chromosomes to the two smaller pieces of paper to represent the cells as they appear after telophase I and cytokinesis.

9. Both of these still have to go through meiosis II. There is no further replication of DNA.

10. Mimic prophase II by removing the nuclear envelope. You no longer have homologous chromosomes pairing, and there is no synapsis or crossing-over at this point.

11. Metaphase II looks much like metaphase of mitosis. The chromosomes align at the cell equator with their centromeres on the equatorial plane.

12. During anaphase II the sister chromatids, now daughter chromosomes, separate as the centromere disappears. They move toward opposite poles of the cell.

13. By the end of telophase II and cytokinesis you have four cells, each with a haploid number of chromosomes.

Study Guide

- Be able to define the terms in bold.
- Be able to describe the processes of mitosis and meiosis.
- Be able to compare meiosis I with mitosis.
- Be able to compare meiosis II with mitosis.

Conclusions

1. Explain the difference between chromosomes and chromatids.

2. When in the life of a cell is DNA replicated? What is the implication of this on cell division?

3. Why is cytokinesis technically not part of mitosis?

4. Describe the role of the spindle fibers in mitosis and meiosis.

5. Describe the key events in each of the phases of mitosis.
   a. Prophase

   b. Metaphase

   c. Anaphase

   d. Telophase

6. Why is crossing-over important to the survival of a population?

7. Describe the products of mitosis with respect to the number of cells produced and their genetic complement.

8. Why is meiosis an essential process in a sexually reproducing species?

Chapter 4

Plant Tissues and Herbaceous Stems

Laboratory Activities

Activity 4.1: Fundamental Tissues

Activity 4.2: Surface Tissues

Activity 4.3: Herbaceous Stems

Activity 4.4: Longitudinal Growth of the Herbaceous Stem

Goals

Following this exercise students should be able to

- Recognize the three primary types of plant tissues.
- Differentiate between simple and complex tissues and identify examples of each.
- Label the tissues in an herbaceous stem.
- Describe the differences between herbaceous monocot and herbaceous eudicot stems.

Introduction

Mitosis results in the production of new cells that are identical to the parent cell. This process occurs in designated locations within the plant's body. These are specialized tissues called **meristems**. Meristematic cells are responsible for the growth of the plant. There are two types of meristems: primary and secondary. Primary meristems are responsible for an increase in the length of the plant, referred to as **primary growth**. They are found in the tips of stems and roots. Secondary, or lateral, meristems are responsible for the increase in the girth of the plant and produce the tissues we know as wood and bark. This is called **secondary growth**.

The mature body of an herbaceous plant is composed of primary tissues and includes surface, ground, vascular, and meristematic tissues. Each of these tissues is composed of numerous cell types; therefore, these are complex tissues. The cells comprising these complex tissues are one of three fundamental types: parenchyma, collenchyma, and sclerenchyma. Parenchyma and collenchyma can be found as **simple tissues**, composed of only parenchyma or collenchyma cells, respectively, or as parts of **complex tissues**. To avoid confusion with these tissues, if a question refers to a "tissue type," the answer would be one of these three fundamental tissues. If a question refers to a specific tissue, that tissue will have a more precise name and the fundamental tissues would not be an appropriate answer.

In this lab, you will be examining the three fundamental tissues as well as examples of complex tissues.

Activities

■ Activity 4.1: Fundamental Tissues

Plants have three fundamental types of tissues. Each tissue is composed of multiple cells to accomplish a particular function. The cells within a tissue may be all of one type or of multiple types. A simple tissue is composed of only one type of cell, whereas a complex tissue is composed of multiple cell types. The three fundamental tissues of plants are **parenchyma**, **collenchyma**, and **sclerenchyma**. Parenchyma and collenchyma can be found as simple tissues, composed of parenchyma and collenchyma cells, respectively. Sclerenchyma cells are typically found embedded within tissues containing other cell types.

Parenchyma cells have thin cell walls and can be somewhat irregular in shape. They often have large vacuoles and function in storage of materials. They can be found in virtually any part of the plant and are the most common cell type in a plant. Collenchyma cells have walls of irregular thickness, with thin cell walls around a portion of the cell and thicker regions of cell wall deposits at the corners of the cells. They offer flexible support to the plant and form the main supportive tissue in herbaceous plants. Sclerenchyma cells have evenly thick cell walls surrounding the cell. There are two common types of sclerenchyma. **Fibers** are elongated cells that often support internal structures of the plant. Fibers are often used to make cloth; cotton is an excellent example. **Sclereids**, also called stone cells, are sclerenchyma cells that are roughly box-shaped and usually found embedded within other tissues, often parenchyma.

Sclerenchyma cells form thick cell walls; sclereids have thicker cell walls than fibers. During wall formation the sclereid cells remain metabolically active and use plasmodesmata to communicate with one another. These plasmodesmata extend through the cell walls in areas called pit canals. The **pit canals** connect the inner lumen, an open area that originally housed the cytoplasm, to the outer edge of the cell wall. These pit canals can often be seen as small, darker bands extending from the innermost part of the sclereid outward.

A typical plant organ is composed of a combination of these fundamental tissues as well as specialized complex tissues. Plant organs include stems, leaves, roots, and flowers, each of which is examined in this chapter.

Initially, the cells produced by a meristem are parenchyma. Some remain thin-walled parenchyma, whereas others begin to change into other cell types. Those that mature into other types of cells may form collenchyma, fibers, sclereids, or one of the other cell types present in the mature plant body.

1. Make a wet mount of a cross-section of celery petiole. The petiole is a structure that connects the flat part of a leaf to the stem. In celery, it is the part we eat. Remember, the thinner the section for a wet mount, the better you'll be able to see the cells under the microscope.

2. Observe the celery petiole under low power. Just inside the epidermis you will see some clusters of light gray cells. These are collenchyma cells. They comprise the celery "strings."

3. Observe the cells in the middle of the petiole. They are larger in diameter and have thinner cell walls than the collenchyma near the edges. These are parenchyma cells. Note the intercellular spaces between the cell walls and the variety of shapes that are present.

4. In the space provided, draw three parenchyma cells and three collenchyma cells. Label the **cell wall** and **cytoplasm** of each. Be sure to also label the cell types in each drawing.

5. Prepare a wet mount of a small portion of the fleshy part of a pear. The fleshy parts of fruits primarily contain parenchyma; however, the pear also contains sclereids. These stone cells occur in clusters that appear as large, gray structures that can vary in shape. Each sclereid is roughly box-shaped with evenly thick cell walls. In these cell walls you may see some thin lines extending from the innermost part of the cell, the **lumen**, toward the outer edge of the cell wall. These are **pit canals** that housed plasmodesmata when the cell was immature.

6. If your sample has large clusters of sclereids and few individuals, remove the slide from the microscope and gently press down on the surface of the coverslip. You may use the eraser end of your pencil for this. The pressure will help separate the sclereids in the clusters.

7. Observe the slide under low power. Locate an individual sclereid or one that is attached to only one or two other sclereids. Sketch them below. Label the cell wall, lumen, and pit canal.

COMPARE AND CONTRAST

1. Describe two ways in which sclerenchyma differs from parenchyma.

2. Identify the characteristic(s) of collenchyma that set it apart from either parenchyma or sclerenchyma.

....

....

....

....

3. Fibers and sclereids are both types of sclerenchyma cells. How are they similar to one another? How are they different from one another?

....

....

....

....

....

■ Activity 4.2: Surface Tissues

Each organ contains permanent tissues in addition to the three primary tissue types that may be present. These tissues are complex in structure, often including multiple types of cells. Most plant organs are covered by **epidermis**. The epidermis is typically composed of one layer of cells, most of which are epidermal cells. In this layer of epidermal cells are some specialized pores call **stomata** (singular = stoma). Each stoma is surrounded by two sickle-shaped cells called guard cells. The **guard cells** are the only cells in the epidermis to contain chloroplasts. Their function is to regulate the opening and closing of each stoma. Stomata are necessary to allow gas exchange with the interior of the plant. Carbon dioxide from the atmosphere must enter through the stoma for photosynthesis. Other gases are also exchanged, including oxygen and water vapor that are typically lost from the interior of the plant through open stomata.

The epidermis is found as the outer covering of leaves, flowers, herbaceous stems, and nonwoody roots. The cell walls of the epidermal cells remain thin and flexible. A waxy substance called **cutin** is secreted by these epidermal cells and forms a layer on the outer surface of the epidermis, called the **cuticle**. Technically, the cuticle is part of the epidermis; however, the cuticle is acellular. The epidermis serves a number of functions, the most important of which is to prevent desiccation. Modifications of the epidermis in some plants can provide additional protection against excessive sunlight and herbivory. One common modification of the epidermis is the presence of **trichomes**. Trichomes are short hairs that grow off of the epidermal cells. The epidermal cells may secrete a number of substances that make the plant less palatable to herbivores.

Woody stems and roots have a stronger, more protective outer covering comprised of a tissue called **periderm**. Periderm consists of primarily cork cells. Recall that cork cells are dead at maturity and have thick cell walls. The thicker cell wall makes a layer of cork cells more protective to the plant than a layer of epidermal cells. The periderm also contains cells called **phelloderm**, which are similar to cork cells but have thinner cell walls and are produced interior to the cork cells, and the meristem that produces cork and phelloderm cells, the **cork cambium**.

1. Observe a wet mount of cork cells. What is the typical shape of a cork cell?

....

....

....

2. Describe the cell wall of the cork cells.

....

....

....

3. Obtain a leaf from *Zebrina*. Tear off a small section of the lower side of the leaf. Try to get a section where the lower epidermis has been separated from the interior cells of the leaf. Make a wet mount. Be sure to orient the leaf section with the *lower* surface facing upward.

4. Observe the wet mount under low power. Locate areas that are light in color. Sometimes they will have a green-colored tint. Center one such area in your field of view and observe under high power. You should be able to see a **stomatal apparatus** consisting of two guard cells on either side of a stoma.

5. In the space below, diagram the stomatal apparatus. Label the stoma and a guard cell.

COMPARE AND CONTRAST

1. Describe two ways in which the periderm differs from the epidermis.

....

....

....

....

....

2. Based on what you know about the fundamental tissues, what type of tissue is the epidermis? What type of tissue is the periderm?

3. Stomata are important for carbon dioxide exchange. Why do you believe they are absent from the periderm?

■ Activity 4.3: Herbaceous Stems

An herbaceous stem is one that typically remains green and flexible. It does not contain wood or bark. As a result the surface is covered by an epidermis and may have trichomes present. The cuticle is usually thin over the surface of the epidermal cells. Stomata and guard cells are present, though in lower numbers than found in leaves. Just inside the epidermis is a region of parenchyma, typically with some collenchyma near the outer edge of the stem, which is called the cortex. The cortex of a stem is usually involved in storage of food reserves. Some of these parenchyma cells contain numerous chloroplasts. These are usually near the epidermis where light levels are higher. Parenchyma cells that contain large numbers of chloroplasts are called **chlorenchyma**. Their primary function is photosynthesis. Although photosynthesis is normally associated with leaves, it is important to recognize that stems can also be major contributors to the overall productivity of a plant. In fact, for some plants the stem is the only site of photosynthesis.

Interior to the cortex are the **vascular tissues**. In an herbaceous stem the vascular tissues are usually organized into bundles. These bundles contain both **xylem** and **phloem** and extend vertically throughout the plant's body. The spatial arrangement of the bundles within the stem differs significantly between monocots and eudicots. In monocot stems the vascular bundles are found scattered throughout the stem's interior. They are more numerous toward the outer edge, but there is no definite patter to the locations of the bundles. In an herbaceous eudicot stem the bundles are arranged in a circular pattern. This pattern provides a clear line with which to distinguish the cortex from the pith. The pith is found in the center of a eudicot stem and is composed of parenchyma cells. Some parenchyma cells extend outward from the pith toward the cortex, filling in the space between the vascular bundles. These are called **pith**, or medullary, **rays**.

Because monocots do not have a circular pattern of vascular tissues, the term "ground tissue" is used to refer to all the parenchyma within the cortex. The cortex extends from the epidermis inward to the first vascular bundles.

GROUND TISSUES

Every plant organ is composed largely of **ground tissues**. These tissues perform many varied functions from storage of water or food reserves to photosynthesis. Most cells in ground tissues are parenchyma, although some collenchyma or sclerenchyma may also be found. The ground tissues can take on many different names, depending on the plant structure involved and the location of the tissues within that structure. Some common types of ground tissues are pith, found in the center of some roots and stems; cortex, found near the edge of roots and stems; and mesophyll, found in leaves. You will see examples of all of these types of ground tissues as you proceed through this chapter.

VASCULAR TISSUES

Vascular tissues are responsible for the transportation of water, minerals, and sugars from photosynthesis throughout the plant's body. Plants have two types of vascular tissues: xylem and phloem. **Xylem** is responsible for water and mineral transportation and is composed of transport cells, vessels, and tracheids as well as parenchyma and fibers. Xylem produced by a primary meristem is referred to as **primary xylem**. Herbaceous stems typically contain primary xylem with no secondary xylem (produced from secondary meristems such as the vascular cambium).

Vertical transport of water is mediated by **vessel elements**. Each vessel element is a cylindrical cell with a thick cell wall. They are dead at maturity. The end of one cell meets the end of the adjacent cell. The end plates of these vessel elements contain numerous pores that allow free movement of water from one cell to the next. Vessels also have perforations in the side walls that allow water to move between adjacent cells laterally (**Figure 4-1**).

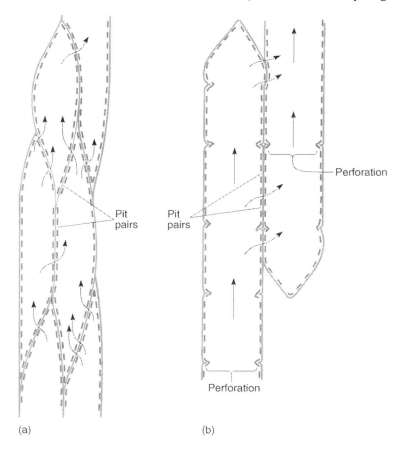

(a) (b)

FIGURE 4-1 Water flow in (a) tracheids and (b) vessel elements. (© Jones & Bartlett, LLC)

Thus, vessel elements in a series effectively create a network of tubes extending vertically through the plant. **Tracheids** are similar in structure; however, they possess tapered ends that overlap one another and are smaller in diameter. Tracheids also have lateral pit pairs that allow water to move from one cell to the adjacent cell laterally as well as vertically (Figure 4-1). The parenchyma cells in xylem are often found intermixed with the vessel elements and tracheids. Xylem parenchyma is responsible for lateral transport of water and minerals.

Phloem transports dissolved sugars throughout the plant in sieve tubes, composed of cells called sieve tube members. The **sieve tube members** have no nuclei at maturity and are closely associated with companion cells to compensate for the absence of internal nuclear control. Phloem also contains parenchyma and often fibers to help support the sieve tubes. Phloem parenchyma functions in lateral transport of dissolved sugars and can be found intermixed with sieve tube members and companion cells. The fibers associated with phloem are usually found in a bundle just outside of the conducting portion of the phloem. Phloem produced by a primary meristem, as in herbaceous stems, is called **primary phloem**.

OBSERVING A MONOCOT STEM

1. Obtain a slide of a cross-section of *Zea mays* (corn) stem.
2. Observe the stem cross-section under scanning power. At this magnification you should be able to see vascular bundles within the stem. In the space below, diagram the location of the vascular bundles within the stem. You do not need to draw every cell; just provide the relative locations of the vascular bundles to one another.

3. Observe the stem cross-section under low power. The cells should be more visible. Move the slide so a vascular bundle is centered in your field of view. Observe the vascular bundle under high power.
4. The vascular bundle of monocots, such as corn, is rather complex. In each bundle the primary xylem is found toward the interior of the stem, whereas the primary phloem is found toward the exterior of the stem. Xylem can be identified by locating vessel elements. These vessel elements are the largest in diameter of all the cells within the bundle and have a thick secondary cell wall that typically stains red. The primary xylem also contains parenchyma cells adjacent to the vessel elements and may contain an air space that often appears oval in shape and can be confused with a vessel element. The air space is interior to the vessel elements (toward the center of the stem). Exterior to the xylem, you will see the primary phloem, which is characterized by two types of cells: the larger sieve tube members (much smaller than vessel elements) and the smaller more darkly colored companion cells. If you think of this vascular bundle as a face, the primary phloem is the forehead, the eyes are the vessel elements, and the air space is the open mouth.

5. On **Figure 4-2**, label the primary phloem, primary xylem, vessel element, sieve tube member, companion cell, air space, and ground tissue.

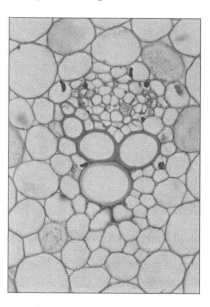

FIGURE 4-2   High magnification of a corn vascular bundle (×250). (Courtesy of James D. Mauseth)

OBSERVING A EUDICOT STEM

1. Obtain a prepared slide of a cross-section of *Ranunculus* (buttercup) stem.
2. Observe the slide under scanning power. Note the location of the vascular bundles and sketch them in the space below. Again, do not include all the cells. Simply diagram the relative locations and sizes of the vascular bundles.

3. Move the slide so a vascular bundle is in the center of your field of view. Observe the vascular bundle under low power. High power may be used if needed.
4. The vascular bundle of an herbaceous eudicot stem has the same general pattern of tissues as does a monocot stem. The primary phloem is toward the exterior of the stem, whereas the primary xylem is toward the interior of the stem. The large vessel elements are still the largest diameter cells inside the vascular bundle and provide an easy marker by which to identify the primary xylem. You should be able to see the sieve tube members as well as companion cells in the phloem. The phloem is covered by a cap of fibers. These phloem fibers provide additional support and protection for the vascular tissues.

5. On **Figure 4-3**, label a vascular bundle, pith, cortex, primary xylem, primary phloem, and phloem fibers.

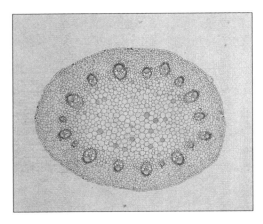

FIGURE 4-3   Cross-section of *Ranunculus* stem (×20). (Courtesy of James D. Mauseth)

6. Obtain a prepared slide of a cross-section of *Helianthus* (sunflower) stem. This is another example of an herbaceous dicot.

7. Observe a vascular bundle under low power. High power may be used if needed.

8. The vascular tissues within the *Helianthus* stem are more organized within each vascular bundle than what was seen with the *Ranunculus* stem. The cap of phloem fibers is larger and more easily distinguished. The primary xylem is arranged with layers of parenchyma in between the layers of vessel elements. The primary phloem is found between the primary xylem and the phloem fibers.

9. On **Figure 4-4**, label the primary xylem, primary phloem, phloem fibers, a vessel element, and a pith ray.

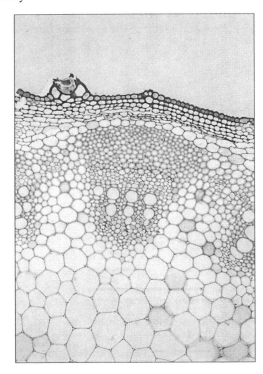

FIGURE 4-4   Vascular bundle of *Helianthus* (×60). (Courtesy of James D. Mauseth)

COMPARE AND CONTRAST

1. Based on location within a vascular bundle, how can you locate xylem versus phloem?

   ....

   ....

   ....

   ....

   ....

2. Based on cell structure, how can you differentiate between xylem and phloem in an herbaceous stem?

   ....

   ....

   ....

   ....

   ....

3. Describe three differences relating to vascular bundles between herbaceous eudicot stems and monocot stems.

   ....

   ....

   ....

   ....

   ....

4. How are the pith and cortex similar? How are they different?

   ....

   ....

   ....

   ....

   ....

■ Activity 4.4: Longitudinal Growth of the Herbaceous Stem

Recall that plants grow at meristems. Stems have an **apical meristem** found at the tip. It is responsible for an increase in the length of the stem as new cells are produced.

The cells produced by the apical meristem begin as parenchyma cells that soon begin to differentiate. Three tissue regions form from these original cells: the **protoderm**, the **ground meristem**, and the **provascular tissue**. The protoderm forms the outer covering of the embryonic stem. As the stem matures the protoderm forms the epidermis. The ground meristem is found in the center of the stem and just inside the protoderm. It is responsible for producing ground tissues, including the pith and cortex. The earliest vascular tissue formed is composed of **protoxylem** and **protophloem**. Collectively, these cells are referred to as **provascular tissues**. They also include meristematic cells that produce additional xylem and phloem; thus, the term procambium is sometimes used to refer to these early vascular cells.

These tissues are found in an embryonic region of the stem called the **subapical meristem**. They each represent an early stage of development. For example, the protoderm cells form a precursor to the epidermis. At this stage of development they are not true epidermal cells, but as they mature they will become the epidermis. Likewise, as the stem matures, protoxylem is replaced with metaxylem and protophloem with metaphloem. Similarly, a vascular cambium may develop in some plants and will produce the secondary xylem and secondary phloem characteristic of plants with secondary growth and woody bodies.

There are some unique external features on a growing stem tip as well. Leaves begin as **primordia**, which often look very different from a mature leaf. These leaf primordia are found adjacent to the apical meristem. The **apical meristem** is found at the tip of the stem and can be recognized by the numerous small cells packed densely together. The primordia are made up of the embryonic tissues: protoderm, ground meristem, and provascular tissues. As the leaves mature these tissues are replaced with the permanent tissues: epidermis, pith, cortex, xylem, and phloem. Once the mature tissues are present, the leaf itself will be mature and is no longer called a primordium. **Buds** are specialized structures on the exterior of a stem that house meristems. The apical bud contains the apical meristem. In a young, growing stem tip the apical bud is difficult to see, although the meristem can be identified easily. Typically, each leaf will have a bud at its base. These are called **axillary buds** and are found in the space between the leaf and the stem, called the **axil**.

1. Obtain a longitudinal section of a *Coleus* stem tip.
2. Observe the slide under low power. Identify the structures discussed on your slide.
3. On **Figure 4-5**, label the apical meristem, leaf primordium, provascular tissue, ground meristem, protoderm, and axillary buds.

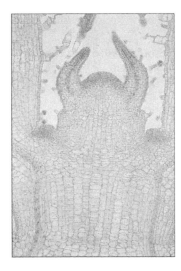

FIGURE 4-5 Longitudinal section through *Coleus* shoot tip (×100). (Courtesy of James D. Mauseth)

4. Look at the region of the stem below the apical meristem and leaves. This region contains mature tissues. Sketch a small section of this mature stem region below and label the epidermis, vascular tissue, pith, and cortex.

5. Observe the region of the specimen where a leaf is attached to the stem. You should be able to see the vascular tissues in the stem. The vascular tissue extends vertically throughout the stem; however, when it reaches a leaf, part of the vascular tissues branch off and enter a leaf. This is called a **leaf trace**. Draw a leaf trace from your specimen below.

6. If you imagine that stem without a leaf attached to it, the vascular tissues would continue in a vertical column. Because the leaf trace has formed, there is a small gap in the vascular tissues. This is called the **leaf gap**. It can be found on the upper side (toward the stem tip) of the leaf trace. Label the leaf gap on your drawing above.

COMPARE AND CONTRAST

1. Describe the similarities and differences between the epidermis and the protoderm.

....

....

....

....

....

2. Describe the similarities and differences between the ground meristem and the pith and cortex.

3. Describe the similarities and differences between the provascular tissues and the xylem and phloem.

4. How does a leaf primordium differ from a mature leaf?

Study Guide

- Be able to define the terms in bold.
- Be able to label the structures identified in the drawings.
- Be able to analyze the differences between parenchyma, collenchyma, and sclerenchyma and recognize examples of each.
- Be able to describe the differences between monocot and eudicot stems.
- Be able to answer each of the questions asked in the lab exercise.

Conclusions

1. List one function for each of the fundamental tissues of a plant.

2. Which is the most common type of fundamental tissue found in a plant?

3. What tissue is found in the center of a eudicot stem? What type of tissue is this?

4. Compare and contrast pith and cortex.

5. Why do we not use the term pith to refer to a tissue in a monocot stem?

6. Describe two differences between the stems of monocots and herbaceous dicots.

7. Describe the relative locations of primary xylem and primary phloem within a vascular bundle.

8. List the names and functions of two cell types found in primary phloem and two cell types found in primary xylem.

9. Where can you find the apical meristem of a stem?

10. In what stem region can you find the protoderm, ground meristem, and provascular tissues?

Chapter 5

Leaves

Laboratory Activities

Goals

Following this exercise students should be able to

- Identify the tissues of a leaf.
- Recognize specialized types of leaves and understand their functions.
- Understand the differences between monocot and eudicot leaves.
- Describe the functions of leaves.

Introduction

Leaves have numerous functions, including photosynthesis, protection, and support. The most familiar of these are foliage leaves that are primarily responsible for photosynthesis. When you think of a leaf, many of you think of foliage leaves that are green and photosynthetic. Although this is the most common type of leaf, many specialized leaves can be found in the plant kingdom as well. Some leaves, such as those of cacti, are modified into **spines**. They are not green and do not function in photosynthesis; however, they do provide a high level of protection against herbivory. In this lab you will be introduced to a number of specialized leaves.

Foliage leaves are composed of the same types of tissues seen in herbaceous stems. Some have slightly different names in leaves as opposed to stems, but all three fundamental tissues may be present as well as the complex tissues found in other herbaceous plant organs. These complex tissues include the epidermis, xylem, phloem, and the ground tissue of the leaf, called the **mesophyll**. In leaves the vascular bundles are called **veins** and are continuous with the vascular bundles of the leaf trace. They appear as raised ridges on the surface of a leaf.

Activities

▪ Activity 5.1: External Features of a Monocot Leaf

Leaves of monocots are distinct from those of eudicots in several ways. Monocots typically possess long, tapered leaves. These leaves are usually sessile, meaning the flattened leaf blade connects directly to the stem. The region of a stem where a leaf, or bud, attaches is called a **node**. The stem region between nodes, where no leaves are attached, is called the **internode**. The venation, of pattern of veins in a leaf, also varies between monocots and eudicots. Pay close attention to this as you examine the specimens in this lab.

1. Observe a leaf from a preserved or living monocot. Some examples may include grasses, irises, or tulips. What species are you observing?

2. You should be able to see veins easily. Sometimes the veins are more pronounced on the lower surface of the leaf than on the upper surface. Describe the pattern of veins present in this specimen.

3. The veins of a monocot are in a parallel pattern. They all enter at the lower end of the leaf and extend upward toward the tip. They do not branch.
4. Monocot leaves have **sheathing leaf bases** that extended around the stem and attach the leaf to the stem.
5. Sketch a monocot leaf below. Label the sheathing leaf base and a vein.

▪ Activity 5.2: External Features of Eudicot Leaves

Eudicot leaves are typically more complex than those of monocots. They are usually divided into a flattened blade, or **lamina**, where most of the photosynthesis occurs and a stalk-like **petiole** that attaches the blade to the stem. At the base of the petiole, paired stipules may be present. Stipules can appear green and leaf-like, flat and scale-like, or may be thin and tapered like spines. Some plants do not have stipules at all, but when present they are always found in pairs.

The veins in a eudicot leaf are often highly branched, forming a pattern called **reticulate** venation. Reticulate veins are found in two common forms. Pinnate veins are characterized by one midvein extending down the length of the lamina. Smaller veins branch off this midvein and continue to branch and become smaller as they near the edge, or **margin**, of the lamina. It is typically encased within a midrib. Palmate veins are characterized by several major veins entering the lamina. Smaller veins branch off these major veins and continue to branch until they reach the margin. There is no midvein or midrib in a leaf with palmate veins.

The margins of leaves vary dramatically and are species specific. An **entire** margin has a smooth, even edge. **Serrate** margins have toothed patterns that give the leaves a jagged edge. Divided margins are characterized by deep invaginations of the leaf surface. Typically, a divided leaf has three to five lobes separated by areas with little or no lamina present. Numerous other forms also exist. Several of the more common types of leaf margins are depicted in **Figure 5-1**.

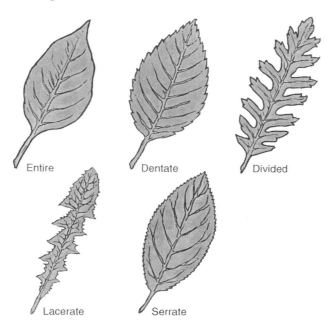

Entire   Dentate   Divided

Lacerate   Serrate

FIGURE 5-1 Several common types of leaf margins. In nature, hundreds of types exist, many intermediate between two or among several other types. (© Jones & Bartlett, LLC)

When a single leaf is attached to a node, it is indicative of an **alternate** arrangement. Usually, the leaves are attached on alternate sides of the stem as you look from one node to the next, and there is never more than one leaf per node. In situations where there are two leaves attached at a single node, leaves are attached in an **opposite** pattern. Some plants have three or more leaves attached at one node. This is a **whorled** pattern. The pattern of leaf attachment to a stem is known as phyllotaxy.

Simple leaves are those with a lamina that is not subdivided into leaflets. The leaflets of a compound leaf look like small leaves. Many have a petiole-like structure called the petiolule that supports the blade. The leaflets may be attached to the end of the petiole, forming a palmately compound leaf. Leaflets may also be attached to a linear extension of the petiole called the rachis. This is characteristic of pinnately compound leaves. Leaves have a bud at their bases. This is the axillary bud, found in the axil of the leaf. The **axil** is the angle between the base of the leaf and the stem. Locating the axillary buds is one of the easiest ways to determine whether a leaf is simple or compound.

1. In the space below, sketch a eudicot leaf. Label the lamina, petiole, and a vein.

2. Observe the preserved leaf specimens on display. For each specimen, record the following information in **Table 5-1**:
    a. Plant name
    b. Type of leaf (simple, compound)
    c. For a compound leaf record the type (palmate, pinnate)
    d. Type of leaf margin (entire, serrate, lobed)
    e. Venation (parallel, palmate reticulate, pinnate reticulate)
    f. Phyllotaxy (alternate, opposite, whorled)

TABLE 5-1: Leaf Characteristics

Specimen Name	Type of Leaf	Leaf Margin	Venation	Phyllotaxy

COMPARE AND CONTRAST

1. How does a pinnately compound leaf differ from a branch with opposite leaves?

2. Compare the typical pattern of leaf venation with the shape of the leaves. Do you notice a pattern?

3. Of the specimens recorded in Table 5-1, which two have the most similar leaves? How did you determine this? Does this indicate these two plants are closely related?

▪ Activity 5.3: Internal Features of a Monocot Leaf

Leaves have a layer of **epidermis** on their surfaces. Because most foliage leaves are flat, the epidermis on the upper side of the leaf is called the upper epidermis, whereas the lower epidermis is found on the lower side of the leaf. Both the upper and lower epidermises are complex tissues composed of epidermal cells and guard cells surrounding stomata. There are some differences between the upper and lower epidermises in a typical leaf, although the overall function is the same. In most foliage leaves the stomata are more numerous on the lower epidermis than on the upper epidermis. The lower epidermis is more protected from light and typically has a lower temperature. The rate of transpiration, water evaporation through the leaf, is significantly lower here than on the upper, warmer surface. The upper epidermis may produce **trichomes**. Trichomes, short hairs, protect the leaf surface from too much direct sunlight, which can be damaging and drying to the leaf. They can also help protect the leaf from herbivory, especially when present along with chemicals that are acidic or bitter. The epidermis of a typical leaf is covered with a layer of **cuticle**. The cuticle adds an additional layer of protection against desiccation. The only area of leaf epidermis not coated with cuticle is the stomata. In leaves that are adapted to hot, dry environments the cuticle can become quite thick.

Between the upper and lower epidermises there is a thick layer of parenchyma cells. This is the ground tissue of the leaf, called the **mesophyll**. The parenchyma that forms the mesophyll are specialized for photosynthesis and contain large numbers of chloroplasts. They are technically called **chlorenchyma**. Within the middle region of the mesophyll are the **veins**. The veins house the xylem and phloem and are surrounded by a layer of cells called the **bundle sheath**. The main function of the bundle sheath is to prevent excessive water loss from the vascular tissues. In some plants the upper and lower sides of the bundle sheath have a mass of fibers that extend outward from the bundle sheath to the epidermis. This is called the **bundle sheath extension**.

Within the vascular bundle the large vessel elements mark the location of the primary phloem, which is toward the upper surface of the leaf. The primary phloem is below the primary xylem, toward the lower epidermis. The cells contained within the xylem and phloem are the same as in other parts of the plant. Xylem contains vessels and/or tracheids along with parenchyma cells, whereas phloem contains sieve tube members and companion cells.

1. Obtain a prepared slide of a cross-section of *Zea mays* (corn) leaf. Observe the specimen under low power.

2. Note the epidermis present on the upper and lower surface of the leaf. Which surface has a larger number of stomata present?

3. Identify the mesophyll on your specimen. Can you identify any pattern in the arrangement of the mesophyll cells? If so, describe it.

4. Find a vein. Notice that some veins are larger than others. The largest vein, the midvein, is found inside the midrib in the center of the leaf. It is easier to identify the tissues within the midvein than those within the smaller veins; however, each vein contains the same tissues. Where is the xylem relative to the phloem inside the midvein? How can you distinguish between the xylem and phloem visually?

5. In some monocots the epidermis will contain enlarged cells called bulliform cells. These are much larger than the typical epidermal cell and have a specialized function. When water within the leaf becomes limited, these cells contract as they lose water, causing the leaf blade to curl toward the stem. This curling limits the amount of leaf surface exposed to sunlight and is a protective measure against excessive transpiration. Do you see bulliform cells in your specimen? If so, sketch a bulliform next to a regular epidermal cell.

6. The same general tissue pattern is present in other monocot leaves as well. **Figure 5-2** presents a cross-section of a sugar cane (*Saccharum*) leaf. Label the upper epidermis, mesophyll, lower epidermis, xylem, phloem, bundle sheath, and bundle sheath extension.

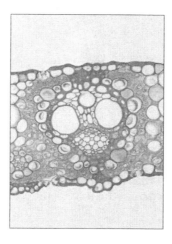

FIGURE 5-2   Vascular bundle in sugar cane leaf (×100). (Courtesy of James D. Mauseth)

COMPARE AND CONTRAST

1. Describe how the lower epidermis differs from the upper epidermis of a monocot leaf.

....

....

....

....

....

2. How does the midvein differ from other veins in the leaf? Describe at least two differences.

....

....

....

....

....

■ Activity 5.4: Internal Features of Eudicot Leaves

Internally, the tissues of a eudicot leaf are extremely similar to those of a monocot leaf. The biggest difference is the structure of the mesophyll. In eudicots the mesophyll is found in two layers. The upper layer consists of rectangular cells that are tightly packed together. This is called the **palisade mesophyll**. The lower layer has large irregular-shaped air spaces between the cells. The cells are more rounded than those in the palisade mesophyll. This lower layer is called the **spongy mesophyll**, because the air spaces give it the appearance of a sponge.

1. Obtain a cross-section of a eudicot leaf, such as lilac (*Syringa*) or maple (*Acer*). Observe the specimen under low power.

2. Observe the upper and lower epidermises. Do you notice any differences between the two?

   ....

   ....

   ....

3. Find the palisade and spongy mesophyll. Where are the air spaces in the spongy mesophyll relative to the stomata in the lower epidermis?

   ....

   ....

   ....

4. Locate the veins. Where are they relative to the palisade and spongy mesophyll?

   ....

   ....

   ....

5. Label the upper epidermis, lower epidermis, cuticle, palisade mesophyll, spongy mesophyll, xylem, phloem, and bundle sheath in the Privet (*Ligustrum*) leaf in **Figure 5-3**.

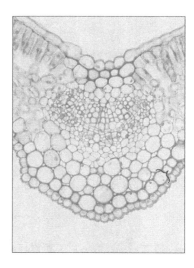

FIGURE 5-3   Transverse section through the midrib of *Ligustrum* (×150). (Courtesy of James D. Mauseth)

COMPARE AND CONTRAST

1. Compare the mesophyll of monocots with that of eudicots.

2. How are the veins of eudicot leaves similar to those of monocots? How are they different?

3. How can you explain the differences between the palisade and spongy mesophyll as they relate to the functions of those tissues?

▣ Activity 5.5: Leaves Adapted to Extreme Environments

Plants live in a variety of habitats. Mesophytes are adapted to moderate levels of moisture and are common throughout much of the world. Xerophytes are adapted to living in dry environments. Hydrophytes are adapted to living in water. Both xerophytes and hydrophytes have many structural adaptations that allow them to survive in their respective environments. The plants observed so far are mesophytes and serve as a comparison to help you identify these adaptations.

Structurally, the leaves of xerophytes exhibit several adaptations, including a thick cuticle, a hypodermis consisting of multiple layers of cells, and stomatal crypts. The thicker cuticle and epidermis decrease the amount of water lost from the leaves by transpiration. The hypodermis is one of several layers of larger cells just below the epidermis that provides an extra layer of protection against desiccation. Having stomata recessed into crypts is also an advantage for conserving water. Air movement in the crypts is limited. With less air moving past the surface of the cells, water molecules that leave the interior of the leaf are not pulled away from the surface of the leaf cells quickly. This limits how much water evaporates through the stomata. Some of those water molecules may reenter the interior of the leaf, further preventing water loss. Xerophytes may have other adaptations in their leaves as well, including trichomes on the upper surface of the leaves to help shade the surface from excessive direct sun.

1. Obtain a prepared slide of a cross-section of a *Ficus* (rubber tree) or *Nerium* (oleander) leaf.

2. In **Figure 5-4** label the cuticle, palisade mesophyll, spongy mesophyll, upper epidermis, lower epidermis, hypodermis, and stomatal crypts.

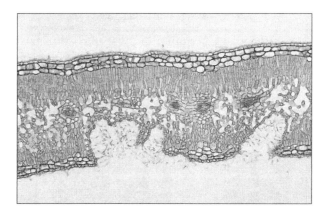

FIGURE 5-4   Cross-section of oleander (*Nerium*) leaves (×80). (Courtesy of James D. Mauseth)

3. Which of the structures labeled above are adaptations to living in dry environments?

....

....

....

The adaptations of hydrophytes, such as water lilies, are quite different from those of the other two groups of plants. These plants must stay afloat in the water column to have sufficient access to sunlight for photosynthesis. Because a water lily lives floating on the surface of still bodies of water, stomata on the lower side of the leaf are not effective for gas exchange. In these plants stomata are found in higher concentrations on the upper epidermis relative to the lower epidermis. There are also numerous large air spaces in the spongy mesophyll. These are much larger than the air spaces of a mesophyte because they help keep the lily afloat. Within the spongy mesophyll there are large, star-shaped sclereids called astrosclereids.

1. Obtain a prepared slide of a cross-section of *Nymphaea* (water lily) leaf.

2. Identify the hypodermis, air spaces, and astrosclereids. How do these air spaces compare with those of the eudicot mesophyte observed earlier?

....

....

....

3. On **Figure 5-5**, label the cuticle, upper epidermis, lower epidermis, palisade mesophyll, spongy mesophyll, a stoma, an air space, and an astrosclereid.

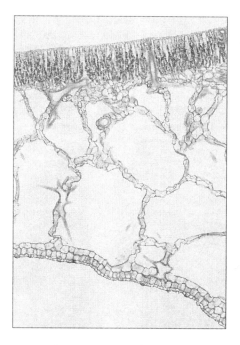

FIGURE 5-5   Cross-section of a water lily leaf (×40). (Courtesy of James D. Mauseth)

COMPARE AND CONTRAST

1. Describe two ways in which xerophytic leaves differ from those of mesophytes.

....

....

....

....

....

2. Describe two ways in which hydrophytic leaves differ from mesophytic leaves.

....

....

....

....

■ Activity 5.6: Specialized Leaves

Leaves can have many functions other than photosynthesis, including support, protection, and storage. Tendrils can be modified stems or leaves. In all cases they function in supporting the plant by wrapping around a solid structure. Spines are leaves that function in protection of the plant. They are thorn-like in appearance and are composed mostly of sclerenchyma. Spines are leaves; thus, they typically have a bud at their bases like foliage leaves. Storage leaves most often are characterized by succulence. Succulent leaves are thick and fleshy. These leaves are green and photosynthetic but have additional parenchyma cells used to store water and nutrients. Succulent leaves often have a thick cuticle, and the plant has other physical and metabolic adaptations to conserving water.

1. Observe the specialized leaves on display.
2. Sketch examples of these leaves below. Be sure to label each sketch with the type of leaf modification as well as the name of the plant, if provided by your instructor.

▪ Activity 5.7: Leaves of Carnivorous Plants

Carnivorous plants are those that obtain nutrients, primarily nitrogen, by digesting insects or other small animals. Nitrogen is an essential element used in the production of many biological molecules, including proteins and DNA. Most plants acquire nitrogen directly from the soil; however, soils vary dramatically in nitrogen content. Some soils, especially swampy areas, tend to have low levels of available nitrogen. The ability of carnivorous plants to acquire nitrogen by other means gives these plants a selective advantage in nitrogen-poor areas.

The leaves of carnivorous plants are modified into traps for catching and dissecting insects. Traps may be passive, with no moving structures, or active. The most familiar of the plants with **active traps** is the **Venus flytrap**. This plant has leaves that form a hinged, two-sided structure. Near the hinge area small hairs act as triggers. Once an insect contacts those triggers, the leaf closes quickly around the insect, trapping it within the sides of the leaf. The leaf then secretes digestive enzymes and slowly digests the trapped insect.

Many other carnivorous plants have **passive traps**. Some have leaves that create fluid-filled pits. Insects that enter the pit, either intentionally or by falling from the upper portions of the leaf, get trapped in the enzyme-rich fluid below. These pits are usually lined with hairs pointing downward to prevent any insects from climbing out. The enzymes in the fluid digest the insects and the cells of the leaf absorb the nutrients. One plant with such a trap is the **pitcher plant**. Another type of passive trap requires the plant to produce a sticky substance on the surface of modified leaves. The sticky substance is produced by the leaves and is exposed on the surface of these leaves. Generally, the sticky fluids appear to be drops of dew on the surface of the plant. Tiny insects that land on the leaves become stuck in the fluid, which also contains digestive enzymes. These traps are not as effective for trapping larger insects but are quite effective at trapping small gnats and flies. **Bladderworts** and **sundews** are two common examples of plants with this type of trap.

1. Observe the carnivorous plants on display.

2. Sketch the leaves of each below. Note whether each is an active or passive trap. Be sure to include the names of the plants as well.

COMPARE AND CONTRAST

1. Consider the appearance of the carnivorous plants you observed. How do the shapes of these plants relate to their mechanism of carnivory?

....

....

....

....

....

2. In comparing all the various leaves you observed today, what are some characteristics common to all of them?

....

....

....

....

Study Guide

- Be able to define the terms in bold.
- Be able to label the structures identified in the figures.
- Be able to analyze the roles of various types of leaves.
- Be able to describe the differences between monocot and eudicot leaves.

Conclusions

1. What type of tissue is the mesophyll?

2. Why are foliage leaves flat?

3. The cells in the epidermis, with the exception of guard cells, are clear and colorless. Why do you believe it is important to the function of the leaf?

4. How can you tell a simple leaf from a compound leaf?

5. What is a rachis? In which kind of leaves can you find a rachis?

6. Why are some plants carnivorous?

7. Where do you find the bundle sheath? Why is the bundle sheath important to the continual function of the leaf?

8. Explain the roles of palisade and spongy mesophyll.

    ....

    ....

    ....

9. List the functions of the leaves you observed today.

    ....

    ....

    ....

10. Describe the difference between parallel and reticulate venation. Which type of plants are associated with each type of venation?

    ....

    ....

    ....

Chapter 6

Roots

Laboratory Activities

Activity 6.1: External Structure of Roots

Activity 6.2: Regions of Growth in a Root

Activity 6.3: Internal Structure of Monocot Roots

Activity 6.4: Internal Structure of Eudicot Roots

Activity 6.5: Growth of Lateral Roots

Activity 6.6: Root Symbioses

Goals

Following this exercise students should be able to

- Identify the key tissues within a root.
- Describe the types of root system.
- Understand the role of the pericycle and why its location is critical.
- Recognize examples of specialized roots.
- Define the roles of plants and symbionts in root nodules and mycorrhizae.

Introduction

Roots serve many functions in plants, including water absorption, anchoring the plant to the soil, and hormone production. The hormones produced by the root system function in part to ensure the growth of the shoot is balanced by the growth of the roots. This is critical to ensuring the root system is sufficient to supply water and minerals to the rest of the plant.

The roots are composed of the same basic cell types found in stems and leaves. Roots of herbaceous plants, or the young roots of woody plants, possess many of the tissues found in stems, including epidermis, cortex, pint, xylem, and phloem. Roots also have some tissues and structures not found in other parts of the plant, such as the endodermis that regulates water entering the xylem.

Roots are **heterotrophic** organs. They are not capable of photosynthesis; thus, they rely on the carbohydrates produced in other parts of the plant to supply their energy needs. This necessitates a direct connection between the vascular tissues in all parts of the plant. Phloem is the tissue that transports carbohydrates throughout the plant. Carbohydrates are loaded into the phloem at sources, areas that are producing the sugars. Sources include photosynthetic structures as well as storage roots, which contain starch that can be broken down into glucose. Xylem also has to be a continuous network throughout the plant to transport water and minerals that enter the roots to other plant tissues. The location and structure of the root vascular tissues facilitates maintenance of this connection of vascular tissues throughout the primary and branch roots.

Activities

■ Activity 6.1: External Structure of Roots

There are two common types of root systems: the **taproot** system and the **fibrous root** system. A taproot system consists of many small lateral roots that branch off a larger, central taproot. The taproot of a mature plant typically extends deeper into the soil than the fibrous root system of an analogous plant. Taproots are usually associated with eudicots and may be woody or herbaceous. Fibrous root systems are composed of a large number of highly branched roots. Some are larger than others; however, there is no large central root present. These roots are derived from the cells at the lower edge of the stem. Because these roots originate from stem cells, they are called **adventitious** roots because they do not originate from other roots.

1. Observe the taproot and fibrous root systems on display.
2. Sketch each one in the space below.

Many other types of roots act in conjunction with these two root systems. Prop roots, commonly found on plants such as corn, are a good example. These are adventitious roots that grow downward from the lower portion of the stem. They are also aerial roots because they are exposed above ground. Prop roots offer an important advantage to corn; they provide additional support to help keep the tall stalks upright.

3. Observe the prop roots on display and sketch an example in the space below.

COMPARE AND CONTRAST

1. How would you describe the main difference between a taproot and a fibrous root system in overall appearance?

2. Corn is a monocot and has a fibrous root system. Why do you believe it needs prop roots for additional support?

■ Activity 6.2: Regions of Growth in a Root

Roots grow from the tip outward, elongating as they move through the soil. The apical meristem is responsible for this growth; however, it is not the outermost structure on a root tip. The **root cap** is a mass of loosely organized parenchyma that covers the growing tip of the root and helps protect the apical meristem from physical abrasion as the root moves through the soil. The apical meristem is located in a cup-shaped region just inside the tip of the root. It comprises the **zone of cell division**. Farther up the root, away from the apical meristem, there is a region where the cells enlarge by absorbing water. This is the **zone of elongation**. Beyond the zone of elongation is the **zone of maturation**. Cells in the zone of maturation are differentiating. At this point all the basic cell types of the herbaceous root are represented: epidermis, cortex, pith, and vascular tissues. The hallmark characteristic of the zone of maturation is the presence of **root hairs**, hair-like extensions off the surface of epidermal cells. Root hairs are part of the epidermal cell and are not multicellular. This zone is responsible for the majority of water absorption in the plant. Moving upward, away from the root tip, beyond the zone of maturation is mature root tissues. At this point the epidermal cells are coated with cuticle and are water resistant. Lateral roots develop in this area as well.

1. Obtain a prepared slide of the longitudinal section of a *Zea mays* (corn) root.
2. Observe the specimen under low power. High power may be needed to observe details of the cells in each region.

3. In the space below, diagram the root tip and label the root cap, zone of cell division, zone of elongation, zone of maturations, and root hairs.

COMPARE AND CONTRAST

1. Describe how cells differ in the zone of maturation and the zone of elongation.

....

....

....

....

....

2. Describe how the cells in the zone of elongation differ from those in the zone of cell division?

....

....

....

....

....

....

■ Activity 6.3: Internal Structure of Monocot Roots

Internally, the structure of roots is somewhat similar to that of stems. Xylem and phloem are the vascular tissues present. Cortex, epidermis, and sometimes pith are present in herbaceous roots. Roots contain some tissues that are not found in stems and leaves. One of the most important is the pericycle, the structure responsible for the growth of branch roots. In addition, the innermost layer of the cortex in a root has a specialized function.

These cells make up the **endodermis**. The endodermis is surrounded by layers of a waxy substance called **suberin**. The suberin coats the upper, lower, and tangential surfaces of the cells forming a water-tight barrier. This barrier is called the **Casparian strip** and serves to regulate the entrance of water and minerals into the xylem.

Monocot roots are characterized by the presence of a central pith. Like the pith of stems, the central pith is composed of parenchyma cells. In monocot roots the pith and the vascular tissues are enclosed within a band of cells that form a cylinder extending vertically through the root. This band of cells is the endodermis, the innermost layer of the cortex, bordered by the Casparian strip. The presence of the Casparian strip causes the endodermis to stain differently from the surrounding cells. On most slides the endodermis appears as a row of cells with thick, red-stained cell walls. All the tissues inside the endodermis are collectively referred to as the **vascular cylinder**, sometimes referred to as the **stele**. Just inside the endodermis is a layer of cells called the **pericycle**. The pericycle may be one-cell thick or may consist of several layers of cells. Pericycle cells maintain their meristematic ability. These cells are the origin of branch roots. The primary xylem can be found in patches near the outer edge of the stele, inside the pericycle. Primary xylem cells are larger in diameter than other cells in the stele. Primary phloem is found in scattered patches more toward the center of the root relative to xylem.

Outside of the stele, the only two tissues present are the epidermis and the cortex. The epidermis is the outer covering of herbaceous roots. In some parts of the root the epidermis contains hair-like extensions called **root hairs**. Root hairs are responsible for most of the water absorption in a root. These are not multicellular hairs like trichomes; instead, each root hair is an extension of a single epidermal cell. The **cortex** occupies the largest amount of space in a root. It extends from the epidermis to the stele, with the endodermis marking the last layer of cells considered to be cortex.

1. Obtain a prepared slide of a cross-section of a monocot root, such as *Zea mays* (corn) or *Smilax* (greenbriar).

2. Observe the specimen on scanning power. You should be able to see the endodermis as a dark circle of cells toward the center of the cell. Everything between that band of cells and the epidermis is the cortex. Approximately what percentage of the root is composed of cortex?

3. Observe the specimen on low power. High power may be used for more detailed observations of tissues. Identify the primary xylem and primary phloem. How can you tell these two tissues apart?

4. On **Figure 6-1**, label the epidermis, endodermis, cortex, pith, primary xylem, primary phloem, and pericycle.

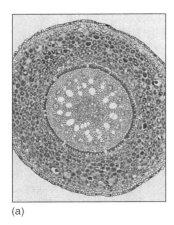
(a)

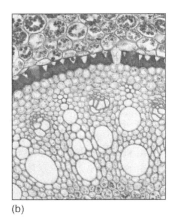
(b)

FIGURE 6-1   Cross-section of *Smilax* root (a, ×15; b, ×150). (Courtesy of James D. Mauseth)

COMPARE AND CONTRAST

1. The cells of the cortex and the pith are all parenchyma. Why do we call them by different names?

2. How does the structure of the epidermis compare with that of the endodermis?

3. How are the functions of the epidermis and the endodermis similar? How are they different?

4. Describe how the cross-section of a monocot root differs from the cross-section of a monocot stem. How can you tell them apart?

....

....

....

....

....

▪ Activity 6.4: Internal Structure of Eudicot Roots

Eudicot roots share many structural similarities with monocot roots; however, the arrangement of tissues within the vascular cylinder varies dramatically. In a eudicot root the tissue in the very center of the root is primary xylem. It is found in a four-pronged arrangement that can be described as the shape of a plus sign (+). Between the arms of the primary xylem are pockets of primary phloem. The primary xylem and primary phloem form the center of the stele. They are surrounded by one or a few layers of cells that form the pericycle. The pericycle is enclosed within the endodermis. Outside of the endodermis is a large layer of cortex with a layer of epidermis forming the outer edge of the root.

1. Obtain a prepared slide of a cross-section of a eudicot root, such as *Ranunculus* (buttercup) or *Helianthus* (sunflower).
2. Observe the specimen on scanning power. Locate the endodermis and the cortex. Approximately what percentage of the root is occupied by the cortex?

....

....

....

3. Observe the specimen with low power. High power may be used to provide greater magnification if needed.
4. On **Figure 6-2**, label the epidermis, cortex, pericycle, endodermis, primary xylem, and primary phloem.

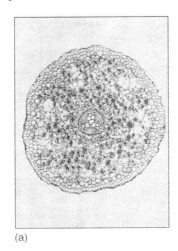

(a)

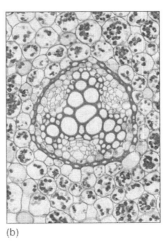

(b)

FIGURE 6-2   Cross-section of *Ranunculus* root (a, ×50; b, ×200). (Courtesy of James D. Mauseth)

COMPARE AND CONTRAST

1. Compare the cortex of monocot and eudicot roots. Which makes up a larger percentage of the root?

2. How are the vascular tissue arrangements in monocot and eudicot roots similar? How are they different?

3. Overall, you should find tissues in the same general order in monocot and herbaceous eudicot roots. List these tissues in order from the outermost to innermost root structures: vascular tissue, endodermis, epidermis, cortex, and pericycle.

4. Describe how a cross-section of a eudicot root differs from the cross-section of a eudicot stem. How can you tell them apart?

▪ Activity 6.5: Growth of Lateral Roots

Lateral (branch) roots grow from the pericycle outward. This location allows the newly formed vascular tissues of the branch root to connect to the vascular tissues of the primary root. As the growing root extends outward it destroys the endodermis, the cortex, and the epidermis of the original root. By the time the new branch root reaches the soil beyond, it will have an intact root cap and apical meristem. It continues to lengthen as it pushes through the soil and develops all the regions of growth seen earlier.

1. Obtain a prepared slide of the cross-section of a *Salix* (willow) root.
2. On **Figure 6-3**, label the primary xylem, primary phloem, endodermis, pericycle, and lateral root.

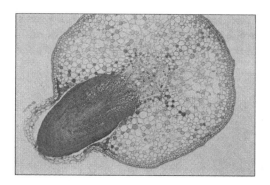

FIGURE 6-3   Lateral root formation in *Salix* (×50). (Courtesy of James D. Mauseth)

COMPARE AND CONTRAST

1. Based on your observations, is *Salix* a monocot or a eudicot? How can you tell?

....

....

....

....

2. How is the structure of the new lateral root similar to the root tip observed earlier? How is it different?

....

....

....

....

....

▪ Activity 6.6: Root Symbioses

A symbiosis occurs when any two individuals of different species live in close association with one another. These relationships can be mutually beneficial (mutualistism), beneficial to one and detrimental to the other (parasitism), or beneficial to one and neither help nor harm the other (commensalism). Plants can participate in a variety of symbiotic relationships; however, we examine two specific examples.

Root nodules form as a mutualistic relationship between plant roots and bacteria. These bacteria, in the genus *Rhizobium*, are capable of nitrogen-fixation that allows them to convert gaseous nitrogen from the atmosphere into ammonia. Ammonia can be used by other organisms, including plants, as a nitrogen source. These bacteria feed on the stored starch inside the root nodules. The plants benefit by acquiring nitrogen from the bacteria, whereas the bacteria benefit by acquiring nutrients and protection within the plant's tissues. **Figure 6-4** shows a cross-section of a root nodule cell. Notice the large number of bacteria within the cytoplasm of a single root nodule cell.

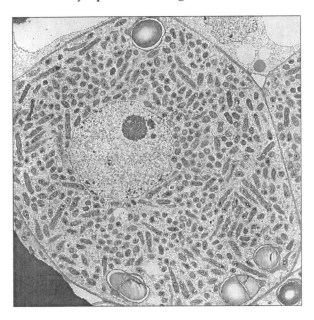

FIGURE 6-4 Bacteria in an infected root nodule cell of cowpea, *Vigna unguiculata* (×5,900). (Courtesy of E. H. Newcomb, University of Wisconsin Department of Botany)

1. Observe the root nodules on display. The bacteria are housed inside the nodules.
2. Diagram a few below.

   Mycorrhizae are associations of fungi with plant roots. The plants benefit from increased water and phosphorus absorption by the fungus. The fungus benefits by obtaining nutrients directly from the plant's tissues. Mycorrhizae appear in two forms. **Ectomycorrhizae** occur when the fungal hyphae (thin filaments of fungal cells) penetrate into the outer layers of root cortex but do not penetrate into the cortex cells. **Endomycorrhizae** occur when the fungal hyphae penetrate deep into the cortex but cannot penetrate the Casparian strip. These filaments enter the cytoplasm of cortex cells without damaging the membrane. Inside the cells they branch continually and occupy a significant portion of the cytoplasm.

1. Obtain a prepared slide of ectomycorrhizae.
2. Observe the specimen under low power. Where do you see most of the fungus?

3. Observe the cortex under high power. How far in to the cortex do the fungal hyphae extend?

4. Obtain a prepared slide of endomycorrhizae.
5. Observe the specimen under low power. Do you see any fungus on the outer surface of the root?

6. Observe the specimen under high power. Can you see hyphae between the cortex cells? Can you see hyphae inside the cells?

7. Approximately how much of the internal space of a single cortex cell is filled with fungal hyphae?

73

COMPARE AND CONTRAST

1. How are the benefits to the plant similar in root nodules and in mycorrhizae?

2. Compare the location and appearance of endomycorrhizae and ectomycorrhizae.

Study Guide

- Be able to define the terms in bold.
- Be able to label the structures identified in the figures.
- Be able to differentiate between monocot and eudicot roots.
- Be able to describe longitudinal root growth and the formation of lateral roots.
- Be able to describe a root symbiosis and provide examples.

Conclusions

1. What is the function of the Casparian strip? What is the primary substance that forms the Casparian strip?

2. Where are root hairs found? What is their primary function?

3. Growth is due to an increase in the number of cells and an increase in the size of cells. Which two regions of plant roots are responsible for root growth?

4. You observe a cross-section of a root and notice the center is filled with parenchyma cells. Is this root from a monocot or a eudicot? What is this central tissue called?

5. What root tissue is responsible for producing lateral roots? Why is it beneficial for the lateral roots to grow from this location?

6. What type of symbiosis is represented by root nodules and bacteria? On what information do you base your answer?

7. What type of symbiosis is represented by mycorrhizae? On what information do you base your answer?

8. Why are grasses and other plants with fibrous root systems preferred for controlling soil erosion?

9. The root cap secretes a slippery substance called mucigel. Based on what you know about the root cap's location, why do you believe mucigel would be useful?

....

....

....

10. Some roots are specialized for other functions, such as storage of water or nutrients. Carrots are excellent examples of nutrient-storing roots. Would you expect the internal structure of a carrot to differ from that of the eudicot root you saw earlier? If so, how would it differ?

....

....

....

Chapter 7

Secondary Meristems and Woody Growth

Laboratory Activities

Activity 7.1: External Growth of Woody Stems

Activity 7.2: Leaf Abscission

Activity 7.3: Internal Structure of Woody Stems

Activity 7.4: Internal Structure of Woody Roots

Goals

Following this exercise students should be able to

- Describe the events of abscission and its importance to woody plant survival.
- Identify key features of woody twigs.
- Compare the internal structure of woody stems and woody roots.
- Describe the tissues found in woody stems and roots.

Introduction

Plants grow in regions of permanent cell growth called meristems. **Primary meristems** increase the length of the plant. **Secondary meristems** increase the girth of the plant and include the vascular cambium, sometimes simply called the cambium, and the **cork cambium**. Both of these secondary meristems are found as cylinders of box-shaped cells running vertically through stems and roots. In both cases, new cells are produced on both sides of the secondary meristem as a result of cell division.

All stems and roots begin as herbaceous structures. In some plants the older portions of stems and roots become woody as epidermis is replaced with cork and secondary xylem is produced in large amounts. Woody parts of the plant provide increased support and protection of internal tissues relative to herbaceous structures.

The vascular cambium produces secondary vascular tissues: **secondary xylem** and **secondary phloem**. Secondary xylem is produced in larger amounts than secondary phloem and forms the tissues we commonly call wood. Secondary phloem is produced on the outer side of the vascular cambium, whereas secondary phloem is produced on the inner side of the vascular cambium, toward the center of the plant.

The cork cambium is found between the secondary phloem and the outer edge of the plant. It produces **cork** cells, or **phellum**, on the outer side of the plant. This forms the outer bark. On the inner side of the cork cambium, **phelloderm** cells are produced. These are similar to cork cells; however, they have a somewhat thinner cell wall. The cork, cork cambium, and phelloderm comprise the **periderm**, the woody outer covering of a stem or root. The term "bark" refers to everything outside of the vascular cambium: the secondary phloem, phelloderm, cork cambium, and cork.

In this exercise we examine some external features of woody stems, as well as the internal structure of woody stems and roots.

Activities

■ Activity 7.1: External Growth of Woody Stems

Woody stems are covered by a layer of cork cells that protect them from abrasion, microbial infection, and desiccation. This tough outer covering also bears a number of scars that tell the story of a particular branch. Every autumn, deciduous trees lose their leaves in a process called **abscission**. This is an ordered process in which the leaf is gradually shut down and sealed off from the stem tissues. Upon completion of abscission, the woody twig bears scars where each leaf was once attached. Called **leaf scars**, they are the same size and shape as the base of the leaf's petiole. Inside the leaf scar are three small circular scars. These are **vascular bundle scars** and mark the locations where leaf traces once enter the leaf. Leaf scars are found at nodes, areas on a stem where leaves and/or buds attach. Often, the **axillary bud** will still be present and the leaf scar will be located directly below it. **Figure 7-1** shows an axillary bud and leaf scar.

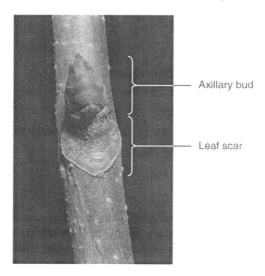

FIGURE 7-1   Woody twig with leaf scar and axillary bud. (Courtesy of James D. Mauseth)

Each stem has multiple buds. The **terminal bud** is located at the tip of the stem, whereas the axillary, or lateral, buds are located at nodes along the length of the stem. Each bud is covered by modified leaves called bud scales that form over dormant buds in the autumn. The bud scales help to protect the meristem found within during winter dormancy. When new growth begins in the spring, the bud scales fall off and the bud produces new growth. This process leaves circular scars encircling the branch. These are bud scale scars. The distance between the current terminal bud and the first set of terminal bud scale scars represents one year's growth.

Cork cells are dead by the time they reach the surface of the plant. Their cell walls accumulate suberin, a waxy substance that serves as a water-proofing agent.

These features help prevent water loss from the surface of the stem or root; however, they also prevent gas exchange between the tissues within the stem or root and the outside environment. This is problematic because plant cells rely primarily on aerobic cellular respiration to furnish needed ATP. To circumvent this problem, woody stems have areas along their outer edges where the cork cells have been disrupted. Thin-walled parenchyma fills the space, and oxygen is able to enter the stem interior. These areas are called **lenticels**.

1. Obtain a woody twig.
2. Identify the terminal bud, axillary bud, leaf scar, bundle scar, bud scale scars, bud scales, and lenticels.
3. Diagram a woody twig and label the structures listed in step 2.

COMPARE AND CONTRAST

1. Compare and contrast the appearance of terminal and axillary buds.

....

....

....

....

....

2. Compare the function of lenticels and stomata.

....

....

....

....

....

▣ Activity 7.2: Leaf Abscission

The process of shutting down leaf activity and removing the leaves from the branches is leaf abscission. One of the first occurrences in this process is an accumulation of suberin in the walls of the cells on the stem side of the petiole. This forms a corky layer that seals off the opening in the stem. On the outer side of this **cork layer** is a **separation layer**. In the separation layer the cells and middle lamellae begin to break down, eventually releasing the leaf from the stem.

1. Obtain a prepared slide of the longitudinal section of a *Populus* abscission zone.
2. Observe the specimen under low power.
3. On **Figure 7-2**, label the abscission zone, vascular bundle, axillary bud, leaf trace, bud scales, and petiole.

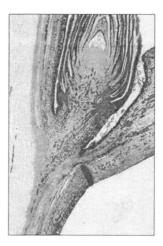

FIGURE 7-2 Longitudinal section through a leaf axil, showing the abscission zone in the petiole and the axillary bud just above (×15). (Courtesy of James D. Mauseth)

COMPARE AND CONTRAST

1. Explain how the abscission zone relates to the scars on a woody twig.

....

....

....

....

....

2. Compare the importance of the plant sealing off the wound formed by abscission with a plant recovering from an insect-induced wound. (Hint: timing is everything!)

....

....

....

....

....

▪ Activity 7.3: Internal Structure of Woody Stems

All woody stems begin as herbaceous stems. The tissues found in the herbaceous stem form the foundation upon which the woody tissues are arranged. At the center of an herbaceous eudicot stem is a pith composed of parenchyma cells and bordered by vascular bundles containing primary xylem and primary phloem. As the stem enlarges a secondary meristem forms between the primary xylem and the primary phloem. This is the vascular cambium. The vascular cambium initially forms within the vascular bundles and subsequently spreads around the circumference of the stem. Once intact, the vascular cambium produces secondary xylem toward the inside of the stem and secondary phloem toward the outside of the stem.

Secondary xylem, or **wood**, is produced throughout the growing season. In the spring, when resources are plentiful, the plant grows quickly and the secondary xylem produced is characterized by large diameter vessels and tracheids. This is called **springwood**. As the growing season continues, resources become limited and growth slows in summer. The cells of the secondary xylem produced during this time are smaller in diameter, which gives them a darker appearance. This is **summerwood**. One year's growth of springwood and summerwood forms an annual ring. **Annual rings** can be used to date the age of a tree and can provide some climate information, particularly concerning rainfall. In dry years trees grow very little, so the annual ring will be much smaller than the annual ring in a wet year.

Throughout the secondary xylem are thin bands of parenchyma that radiate outward from the pit, like the spokes of a wheel. These parenchyma cells function to transport water and minerals laterally, from the secondary xylem outward toward the edge of the stem. These bands of cells are called **xylem rays** or wood rays.

The secondary xylem is produced between the vascular cambium and the primary xylem. The primary xylem remains in place and borders the pith. As the tree continues to grow, sometimes the pith breaks down, forming a pith cavity. Secondary xylem functions in vertical water transport. When cavitation occurs and the water column in one vessel breaks, that vessel will never again transport water. Instead, the nonfunctioning vessel is sealed off and antimicrobial substances are deposited within it. The layers of nonfunctioning xylem are called heartwood. **Heartwood** is found toward the center of the stem and is darker in color than sapwood. **Sapwood** is active xylem found surrounding the heartwood.

The vascular cambium is a thin band of cells that encircles the secondary xylem. It produces secondary phloem toward the outer edge of the stem. Secondary phloem is found in much smaller quantities than secondary xylem. It also stains darker. Within the band of secondary phloem are multiple **phloem rays** composed of parenchyma. Some of these rays are made of thin bands of cells and are called narrow phloem rays. Others are thin near the vascular cambium and branch out into wide rays as they approach the outer bark. The latter are broad (dilated) phloem rays. The broad phloem rays are typically light in color because they do not stain well. The sieve tubes of the secondary phloem generally appear fairly dark due to the stains used in preparing the slide.

Beyond the secondary phloem toward the outer edge of the stem is the thin layer of cells forming the periderm. The periderm is composed of three tissues: phellem (cork), cork cambium (phellogen), and phelloderm. The cork cambium produces both cork and phelloderm. Cork is produced on the outer surface of the stems and forms what is commonly referred to as the outer bark of the plant. Phelloderm is produced on the inner side of the cork cambium. It is difficult to see the cork cambium on prepared slides; however, phelloderm stains lighter than cork because the cell walls are somewhat thinner. Between the periderm and the secondary phloem is a layer of cortex. The term "bark" refers to all tissues outside the vascular cambium, including the secondary phloem, cortex, phelloderm, cork cambium, and cork cells.

1. Obtain a prepared slide of a cross-section of *Tilia* (basswood) stem.
2. Observe the specimen under scanning power. Low or high power may be used to observe more cellular details.
3. Locate the pith. Is it intact or has a pith cavity formed?

....

....

....

4. Locate the secondary xylem. How old was this specimen when it was harvested? How can you tell?

....

....

....

5. Locate the secondary phloem. Approximately what percentage of the secondary phloem is composed of broad phloem rays?

....

....

....

6. Diagram one-fourth of the specimen and label the pith, secondary xylem, secondary phloem, vessel element, xylem ray, phloem ray, vascular cambium, cork, periderm, cortex, and bark.

COMPARE AND CONTRAST

1. Compare the amount of space occupied by secondary phloem with that occupied by secondary xylem.

....

....

....

2. Xylem rays and phloem rays function in lateral transport of fluids. How are their structures similar? How are they different?

■ Activity 7.4: Internal Structure of Woody Roots

Woody roots are arranged in much the same way as woody stems. The periderm provides an outer covering of the root and is composed of a layer of cork on the outer surface, the cork cambium, and phelloderm. The cork cambium arises from the pericycle, and as it produces cork and phelloderm the entire cortex and epidermis are sloughed off of the root. The secondary phloem therefore extends from the edge of the periderm inward to the vascular cambium. Inside the vascular cambium is the secondary xylem. In the center of the root, remnants of the primary xylem can be found.

1. Obtain a prepared slide of the cross-section of a *Tilia* (basswood) root.
2. Observe the specimen under scanning power. Low power may be used if desired.
3. Locate the center of the root. This is where you'll find the primary xylem. Working your way to the outer edge of the root, identify the other tissues.
4. On **Figure 7-3**, label the periderm, secondary phloem, secondary xylem, primary xylem, vascular cambium, xylem ray, and phloem ray.

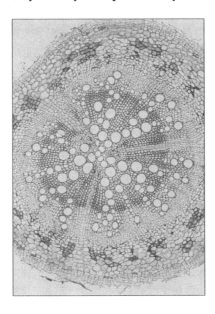

FIGURE 7-3   Cross-section of a woody root (×40). (Courtesy of James D. Mauseth)

COMPARE AND CONTRAST

1. How do the arrangement of tissues in a woody root compare with that of a woody stem?

2. Are growth rings visible in woody roots as well as in woody stems? How can you explain this?

Study Guide

- Be able to define the terms in bold.
- Be able to identify the tissues of a woody stem and woody root.
- Be able to identify the external features of a woody stem.
- Be able to describe the process of leaf abscission and identify the tissues involved.

Conclusions

1. Explain how leaf abscission results in the leaf and bundle scars on a woody twig.

2. Why is there no endodermis in a woody root?

3. How can secondary xylem be used to determine the age of a tree?

4. Why are bud scales important to woody plants?

5. Describe how terminal buds differ from axillary buds.

6. What is the function of rays?

7. Compare the tissues found in the center of woody stems with those found in the center of woody roots.

8. Does Figure 7-3 show a monocot or eudicot root? How can you tell?

9. What main cell types are found in secondary phloem? Are these the same cells found in phloem rays?

10. What is the main function of a lenticel? What is an analogous structure found in leaves?

Chapter 8

Photosynthesis

Laboratory Activities

Activity 8.1: Pigment Chromatography

Activity 8.2: Measuring Photosynthesis

Goals

Following this exercise students should be able to

- Describe the process of photosynthesis in plants.

- Relate the structure of the chloroplast to the procedures of photosynthesis.

- Understand the role of pigments in photosynthesis.

Introduction

Photosynthesis is the process by which light energy is used to convert water and carbon dioxide into carbohydrates. In plants this process occurs in the **chloroplast**. A chloroplast is surrounded by two membranes that encase a fluid-filled **stroma**. Within the stroma are stacks of membranous discs. The discs are called **thylakoids**, composed of thylakoid membranes. Each stack of thylakoids is called a **granum** (plural, grana).

Light is composed of packets of energy, called **photons**, that travel in waves. It is one type of electromagnetic energy that originates from the sun. The electromagnetic spectrum, shown in **Figure 8-1**, depicts the various wavelengths of all types of electromagnetic radiation. Wavelength is defined as the distances between the crests of two adjacent waves. The shorter the wavelength, the more waves will pass a set point in a given amount of time. Thus, radiation with shorter wavelengths contains higher amounts of energy. White light can be divided into its component colors by a prism. The colors range from blue to red in order of increasing wavelength. Each color contains different amounts of energy.

The thylakoid membranes contain chlorophylls and other photosynthetic pigment molecules. A **pigment** is any compound that absorbs a specific wavelength of light. In the case of photosynthetic pigments, the photons absorbed provide energy for photosynthesis. Any wavelengths of light that are not absorbed may be reflected or transmitted by the object containing the pigments. Light that is reflected by an object reaches our eyes and is perceived by our brains as a color. The wavelengths of light reflected therefore determine the colors we see. The colors absorbed are those that cannot be seen.

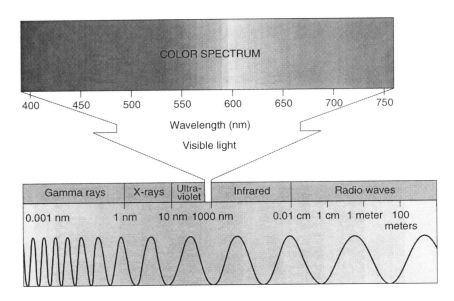

FIGURE 8-1 On the left end of this electromagnetic spectrum are gamma rays with extremely short wavelengths and high-energy levels. As wavelengths become longer, quanta become less energetic. (© Jones & Bartlett, LLC)

Within the thylakoid membranes are clusters of pigment molecules called **photosystems**. These photosystems contain a combination of pigments, including chlorophyll *a*, chlorophyll *b*, and carotenoids. Chlorophyll *a* is the central molecule in photosynthesis. It is responsible for transferring energy from light to electrons. Those energized electrons are passed along a series of molecules through oxidation-reduction reactions. With each transfer the electrons lose some energy, which is used to pump hydrogen ions across the thylakoid membrane into the thylakoid space. In turn, the hydrogen ions power the production of ATP by the enzyme ATP synthase as they diffuse back into the stroma. The electrons are finally transferred to a molecule of NADP$^+$, which is reduced to NADPH. These electrons originate from water molecules that are split, releasing oxygen into the atmosphere and hydrogen ions into the thylakoid space. The products of the light-dependent reactions are ATP and NADPH, both of which are used in the **stroma reactions (Calvin cycle)** to produce energy-rich carbohydrates.

Chlorophyll *b* acts as an accessory pigment, as do the carotenoids. Accessory pigments absorb light at wavelengths other than those absorbed by chlorophyll *a*, thus increasing the amount of usable energy that can be obtained from the light reaching the plant. The energy absorbed by these accessory pigments is passed from pigment to pigment until it reaches one particular molecule of chlorophyll *a* called the **reaction center**. The chlorophyll *a* in the reaction center is responsible for energizing the electrons.

The Calvin cycle is responsible for taking in carbon dioxide from the atmosphere, which enters the leaves through stomata, and reducing it to carbohydrates. This process does not directly require light energy; however, it does require the ATP and NADPH produced from the light reactions. The main product of the Calvin cycle is glyceraldehyde-3-phosphate, a three-carbon sugar molecule that can be used to make glucose or other organic molecules.

The light reactions are the key to the process of photosynthesis because the pigments involved acquire the energy necessary for the remainder of the reactions. Remember, every plant relies on a combination of pigments to harvest this energy. **Chlorophyll *a*** is a yellow-green pigment, whereas **chlorophyll *b*** is a blue-green pigment. These are typically the most common pigments found in green leaves. **Carotenoids** are yellow-orange pigments that are found in leaves year-round, although their color is usually masked by the presence of chlorophyll. These include **carotene**, a darker yellow-orange pigment, and **xanthophyll**, a pale yellow pigment. Other pigments that are not photosynthetic may also be present in leaves. One such pigment is anthocyanin, a pigment that can appear blue or red depending on the pH in the leaf.

Activities

▪ Activity 8.1: Pigment Chromatography

The pigments of the light-dependent reactions are critical to the process of photosynthesis. These can be separated from a mixture of pigments by a process called **chromatography**. In chromatography, substances are separated based on their chemical and physical properties. It involves the use of a stationary phase, which is chromatography paper made of cellulose, and a mobile phase typically composed of an organic solvent.

The process is relatively simple. Solutions of pigments are directly applied to chromatography paper, which is then partially submerged in a solvent. The solvent is absorbed by the paper and travels upward. As it reaches the solution, some substances dissolve in the solvent and travel upward as well. Some substances bind quickly to chromatography paper and remain in place. Pigments can be separated based on how readily they bind to the solvent versus how easily they dissolve in the solvent. The **retention factor** (Rf) can be calculated by dividing the distance traveled by the dye by the distance traveled by the solvent (Rf = distance traveled by dye/distance traveled by solvent).

1. Obtain leaves of spinach, red cabbage, and red kale (or red leaf lettuce).
2. Grind leaves in 95% ethanol using a mortar and pestle for 5 minutes or until liquid is richly pigmented.
3. Obtain three pieces of chromatography paper. Use a *pencil* to draw a line 2 cm from the bottom edge of the paper.
4. Use a **capillary tube** or Pasteur pipette to transfer a small amount of spinach extract to the first piece of paper. Transfer an equal amount of cabbage and kale extract to the other two pieces of chromatography paper.
5. Allow pigment solutions to dry.
6. Repeat steps 3 and 4 four more times for each pigment.
7. Add chromatography solvent (9:1 petroleum ether-to-acetone) to three test tubes to a height of 1.5 cm. Be sure the height of the solvent in the tube does not reach 2 cm!
8. Insert the chromatography paper strips into the test tubes (one piece of paper per test tube) containing the solvent.
9. Observe the migration of the solvent and the pigments. Allow the separation to continue until the solvent (wet line) travels at least three-fourths of the length of the paper. Do not allow the solvent to reach the opposite end of the chromatography paper.
10. Sketch your results below. Include the colors of each pigment and their relative locations.

11. Based on your knowledge of plant pigments, which pigments are likely present in the samples?

12. Which pigment traveled farthest?

13. Record the Rf for each dye in the spinach and cabbage samples in the space below.

14. Based on the Rf factors, can you conclude that any of these dyes found in the cabbage extract are the same as those from the spinach extract. Why or why not?

COMPARE AND CONTRAST

1. How were you able to determine the likely identities of the pigments separated by chromatography?

2. How do the pigments differ between spinach and red cabbage? Between spinach and red kale?

3. What can you determine about the photosynthetic ability of each of these leaf types based on this information?

■ Activity 8.2: Measuring Photosynthesis

Plants need carbon dioxide for photosynthesis. Most leaves in their natural environments acquire this carbon dioxide directly from the atmosphere; however, carbon dioxide can be acquired from water as well. Aquatic plants rely on this as their primary source of carbon dioxide for photosynthesis. Carbon dioxide (CO_2) can combine with water (H_2O) to form carbonic acid (H_2CO_3). In the process excess protons (H^+) are released, lowering the pH of the solution. Therefore, carbon dioxide is an acid even though it does not contain hydrogen.

The pH of a solution can be monitored by using pH indicators, substances that change colors with varying pH. Bromothymol blue and phenol red are two commonly used indicators that change color at approximately neutral (pH 7) conditions. Bromothymol blue is blue under basic conditions, yellow under acidic conditions, and green at a neutral pH. Phenol red is yellow in acidic conditions and pink/red in basic conditions. At a neutral pH it appears orange. Either may be effectively used for analyzing pH during photosynthesis.

1. Add 30 ml of water to six large test tubes. Set one aside for use as a control.
2. Add 1 g of baking soda to each tube.
3. Add three to four drops of a 0.1% solution of phenol red to each of the test tube's contents.
4. Cut sprigs of *Elodea* (anacharis) approximately 1 inch long and place one in each test tube.
5. Cap each test tube or cover with a piece of parafilm.
6. The tubes should be set up as follows:
   a. No baking soda added, not wrapped
   b. Baking soda, wrap in foil
   c. Baking soda, not wrapped
   d. Baking soda, wrap with red cellophane
   e. Baking soda, wrap with blue cellophane
   f. Baking soda, wrap with green cellophane
7. Expose all tubes to light for 30 minutes. Record the color of the liquid in each tube in **Table 8-1**. What color indicates photosynthesis?

TABLE 8-1:	Measuring Photosynthesis Results					
Minutes	Tube A	Tube B	Tube C	Tube D	Tube E	Tube F
0						
2						
4						
6						
8						
10						
12						
15						
18						
21						
25						
30						

COMPARE AND CONTRAST

1. Which treatment exhibited the fastest rate of photosynthesis?

2. Compare the effects of the three colors of cellophane on the rates of photosynthesis. What does this tell you about the colors of light needed for photosynthesis?

3. Compare the results of tubes a and b with those of tube C. Why use tubes A and B?

Study Guide

- Be able to define the terms in bold.
- Be able to describe the importance of pigments in photosynthesis.
- Be able to identify the key pigment and accessory pigments involved in photosynthesis.
- Be able to explain the importance of the light-dependent reactions and the Calvin cycle in the overall process of photosynthesis.
- Be able to explain why water and carbon dioxide are needed in photosynthesis.

Conclusions

1. What molecule is used to produce electrons for the light-dependent reactions? What is the byproduct (waste) of this molecule being split?

2. Explain how the light-dependent reactions are chemically linked to the Calvin cycle.

3. Based on your observations, which leaf was capable of higher amounts of photosynthesis: the spinach, the red cabbage, or the red kale? On what are you basing this conclusion?

4. Describe the stationary and mobile phases of paper chromatography.

5. How can chromatography be used to separate substances?

6. Can chromatography be used to identify substances?

_____

_____

_____

7. Explain how aquatic plants can undergo photosynthesis when they cannot acquire oxygen directly from the air.

_____

_____

_____

8. What pH indicator was used to assay for photosynthesis? Describe how its color relates to the pH of a solution.

_____

_____

_____

9. Why can a pH indicator be used to assay for photosynthesis? What does the color change indicate?

_____

_____

_____

10. Explain the effects of colored cellophane on the rates of photosynthesis based on what you know about light absorbance.

_____

_____

_____

Chapter 9

Cellular Respiration and Fermentation

Laboratory Activities

Activity 9.1: Bacterial Fermentation in the Production of Food

Activity 9.2: Bacterial Fermentation of Sugars

Activity 9.3: Yeasts and Alcoholic Fermentation

Activity 9.4: Cellular Respiration

Goals

Following this exercise students should be able to

- Describe the process of cellular respiration in plants.
- Relate the structure of the mitochondrion to the procedures of cellular respiration.
- Compare processes of cellular respiration and fermentation.
- Explain the importance of cellular respiration to a plant.

Introduction

The ultimate source of energy for plants is sunlight. Energy in sunlight is captured by **photosynthesis** and converted into chemical energy in the form of glucose, other carbohydrates, proteins, and other essential organic molecules. Some of these molecules are subsequently used by plant cells as an energy source to power the internal functions of each cell; however, these initial carbohydrate products cannot directly power a cell's activities. The energy in glucose and other sugars must be converted into a more accessible form. As the sugar molecules are broken down, the energy released is captured and used to produce **ATP** (adenosine triphosphate), the primary energy-carrying molecule in eukaryotic cells. The chemical process responsible for producing most ATP in plant cells, and the cells of many other organisms, is **cellular respiration**.

Cellular respiration is a complex process composed of many chemical reactions. These reactions are divided into three major steps: **glycolysis**, the **Krebs cycle**, and the **electron transport chain**. Cellular respiration in its simplest form begins with the breakdown of glucose. Although other types of organic molecules can serve as energy sources for cellular respiration, glucose is one that goes through the entire series of reactions and thus is the typical example for a starting point in the process.

Glucose is found in the cytoplasm of many plant cells. It is one of the main products from photosynthesis; however, it is typically stored as the polymer starch. Cells can break down starch to release glucose whenever energy is needed. Glucose is initially broken down in the reactions of glycolysis. Glycolysis uses two molecules of ATP to help break glucose, a six-carbon molecule, into two molecules of **pyruvate**, a three-carbon molecule. In the process four molecules of ATP are generated along with two molecules of NADH (reduced form of nicotinamide adenine dinucleotide), an electron carrier.

Glycolysis occurs in the cytoplasm of the cell. As long as sufficient levels of oxygen are present, the pyruvate molecules are shuttled into the matrix of the mitochondria where they are further broken down and enter the Krebs cycle. Pyruvate cannot directly enter the Krebs cycle; it must first be converted to a two-carbon molecule called an acetyl group. This is a preparatory step that occurs within the mitochondrial matrix. To convert pyruvate to an acetyl group, a carbon atom and the atoms bound to it must be removed. That produces one molecule of carbon dioxide. The acetyl group becomes attached to coenzyme A and forms acetyl-coenzyme A (**acetyl-CoA**), which enters the Krebs cycle.

The Krebs cycle continues to break down the carbon-based molecule that originated from the glucose. Because glycolysis produces two molecules of pyruvate per glucose molecule, the preparatory reaction and the Krebs cycle occurs twice for every pass through glycolysis. Acetyl-CoA enters the first reaction of the Krebs cycle when it reacts with the four-carbon oxaloacetate and "drops off" the two-carbon acetyl group. The combination of the two-carbon acetyl group and the four carbon oxaloacetate produces an initial six-carbon product called citrate, or citric acid if dissolved. Thus, the Krebs cycle is often called the **citric acid cycle**. The CoA is recycled and picks up another acetyl group later. The citrate undergoes a series of reactions. Two more molecules of carbon dioxide are produced along with one molecule of ATP, three molecules of NADH, and one molecule of **FADH$_2$** (reduced form of flavin adenine dinucleotide). Because the cycle turns twice, the totals per molecule of glucose double (4 CO$_2$, 2 ATP, 6 NADH, and 2 FADH$_2$). Because of the cyclic nature of the Krebs cycle, the starting product, oxaloacetate, must be regenerated. After the Krebs cycle the carbons from the original glucose have now been converted to CO$_2$. The many electron carriers, NADH and FADH$_2$, from all the earlier reactions can be used in the final stage of cellular respiration: oxidative phosphorylation in the electron transport chain.

The electron transport system is composed of a series of molecules that act as electron acceptors. These molecules are embedded within the inner membrane of the mitochondrion, which is folded into many finger-like regions called **cristae**. The NADH and FADH$_2$ arrive at the inner side of the cristae and release their electrons to the initial electron acceptor of the electron transport system. As the electrons are donated to the molecules of the transport system, protons (H$^+$) are released from the NADH and FADH$_2$. The electrons are passed from molecule to molecule through the electron transport system through a series of oxidation-reduction, or redox, reactions. Oxidation reactions cause the removal of an electron from a molecule, whereas reduction reactions cause an electron to be added to a molecule. In most cases these reactions happen in the same time and at the same place. So, in reality, electrons are removed from one molecule and added onto a neighboring molecule. As these redox reactions continue along the electron transport system, the electrons are moved horizontally along the molecules in the inner mitochondrial membrane. As they pass from one electron acceptor to another, some energy is lost. This energy from the electrons is used to allow some of these electron acceptors to act as proton pumps.

The protons that were released from NADH and FADH$_2$ are then pumped from the matrix across the inner membrane into the intermembrane space. These protons form a gradient with a large amount of protons building up in the intermembrane space. Because protons cannot directly diffuse through biological membranes, they have only one means by which they can diffuse back into the matrix and thus equalize the concentrations of protons on either side of the membrane and that is by passing through a protein called **ATP synthase**. ATP synthase uses the energy from the passing protons to produce ATP from ADP (adenosine diphosphate) and an additional phosphate group. The flow of protons through the ATP synthase molecule is referred to as **chemiosmosis**. Because the energy source for pumping the protons comes from oxidation reactions, this type of ATP production is also called **oxidative phosphorylation**. Up to 34 molecules of ATP are produced in this way. At the end of the electron transport system the electrons are finally combined with **oxygen**, the final electron acceptor, and hydrogen ions (protons) to form **water**.

Although the process of aerobic cellular respiration is complex, it produces an extremely high yield of ATP, most of which is produced in the electron transport system. This is the most complete breakdown of glucose possible in a cell; however, it is not the only means by which usable energy can be extracted from glucose metabolism. Some organisms can use anaerobic respiration, which is similar to the process described except oxygen is not used as the final electron acceptor and some of the chemical reactions differ. In both types of cellular respiration all three major steps, glycolysis, the Krebs cycle, and the electron transport system, occur. Plants are not capable of this type of anaerobic respiration; however, some plant cells are capable of an anaerobic process called **fermentation**.

Fermentation is useful because it does not require the presence of oxygen and can serve as a rapid way for cells in anoxic conditions to generate ATP. The downside of fermentation is that it is much less efficient in producing ATP than is cellular respiration. Most of the energy in the original molecule of glucose remains unused in organic byproducts, usually acids or alcohols.

The process of fermentation begins with glycolysis, which occurs just as it did in cellular respiration. Pyruvate is still the main product of glycolysis; however, it does not enter the mitochondrion or proceed through the remaining reactions of cellular respiration. Instead, the pyruvate undergoes one or two additional reactions in the cytoplasm of the cell. Fermentative plant cells produce ethanol and carbon dioxide from the metabolism of pyruvate. This type of alcoholic fermentation is also seen in yeasts. Some bacteria, fungi, and animal cells produce acids as the byproducts of fermentation, lactic acid being the best known example. Organisms that are capable of both aerobic cellular respiration and anaerobic fermentation are said to be **facultatively anaerobic**. This includes yeasts, some bacteria, and certain plant and animal cells. These facultative anaerobes use cellular respiration as long as oxygen is present. If oxygen is not available, they can survive by fermentation. Other organisms, including some bacteria, are completely fermentative and never go through cellular respiration. These are anaerobic organisms that do not use oxygen at all.

Activities

▉ Activity 9.1: Bacterial Fermentation in the Production of Food

Many bacteria produce some type of acid during the fermentation process. Some species also produce carbon dioxide. Bacterial and fungal fermentation has long been used by humans to produce a number of good products. Bacteria are used to produce a variety of cheeses, yogurt, sauerkraut, and many other fermented foods. In this experiment bacteria is used to produce yogurt.

Many species of bacteria can be used in the production of yogurt; however, the major contributors to yogurt fermentation are *Lactobacillus bulgaricus*, *Lactobacillus acidophilus*, and *Streptococcus thermophilus*. The typical dairy yogurt that is produced commercially contains live cultures of these key bacterial species. As the bacteria grow in the milk, they use the lactose in the milk for an energy source. By metabolizing the lactose they produce lactic acid, which produces a tart flavor in the yogurt.

1. Pour 200 ml of milk into a 1-liter beaker.
2. Using pH test strips or a pH meter, record the pH of the milk below.

3. Heat the milk to 60°C, stirring continually. Maintain that temperature for 10 minutes.
4. Remove the milk from the heat and cool to 45°C.
5. Add 5 g powdered milk to produce a creamier product.
6. Add 1 teaspoon or 5 ml of the starter yogurt culture to the milk. Mix well.
7. Incubate at 37°C for 24 hours.
8. Refrigerate until next lab session.
9. Observe the yogurt. Note any changes in texture, color, smell, and so on below.

10. Using pH test strips or a pH meter, record the pH of the milk below.

11. Based on your understanding of fermentation, how can you explain the results above?

COMPARE AND CONTRAST

1. Contrast facultative anaerobes with aerobes and anaerobes.

2. In what ways did the milk at the beginning of the experiment differ from the final product?

....

....

....

....

....

■ Activity 9.2: Bacterial Fermentation of Sugars

Many bacterial species are capable of fermentation. Some ferment only one or two sugars, whereas others ferment multiple sugars. In this experiment the ability of four species of bacteria to ferment three different sugars is analyzed. Bacterial fermentation is often determined using phenol red fermentation broths. Each broth is a liquid medium containing only one type of sugar: glucose, sucrose, or lactose. **Phenol red**, a pH indicator, is added to the medium to allow for easy detection of acidic byproducts from the fermentation process. At a neutral pH phenol red is a cherry red color. In a basic pH the color sometimes becomes more of a dark pink. At an acidic pH phenol red turns yellow.

   Many bacteria produce acids during fermentation, and some species also produce carbon dioxide. To determine the production of carbon dioxide, each phenol red fermentation broth is in a large-diameter test tube with a smaller tube inverted inside. These inverted small tubes are called **Durham tubes**. They are initially filled with the broth. If carbon dioxide or other gases are produced, they build up inside the Durham tube and appear as a trapped bubble that can become quite large.

1. Obtain broth cultures of *Escherichia coli, Enterobacter aerogenes, Alcaligenes faecalis,* and *Proteus vulgaris.*

2. Obtain one fermentation broth containing each sugar (glucose, sucrose, and lactose) for each bacterium being tested.

3. Inoculate one fermentation broth containing each sugar (three total broths) with 0.5 ml *E. coli* broth culture each.

4. Repeat step 3 for each remaining bacterial species.

5. Incubate at 37°C for 24 to 48 hours. Refrigerate until results can be determined in the following lab period.

6. Record your results in **Table 9-1**.

TABLE 9-1: Results of Bacterial Fermentation							
	Glucose		Sucrose		Lactose		
Organism	Acid	Gas	Acid	Gas	Acid	Gas	Fermentation
Escherichia coli							
Proteus vulgaris							
Alcaligenes faecalis							
Enterobacter aerogenes							

COMPARE AND CONTRAST

1. Compare the fermentation results for *E. coli* and *P. vulgaris*.

   ....

   ....

   ....

   ....

   ....

2. In what ways did the fermentation broths at the beginning of the experiment differ from those with positive results at the end of the experiment?

   ....

   ....

   ....

   ....

   ....

3. How are the products of fermentation seen with *E. coli* different from those seen with *Lactobacillus* in the production of yogurt?

   ....

   ....

   ....

   ....

   ....

■ Activity 9.3: Yeasts and Alcoholic Fermentation

Many types of fungi are capable of fermentation, producing either acidic or alcoholic byproducts. Fungi are typically facultative anaerobes, so fermentation will only be observed if oxygen is absent. Yeasts are unicellular fungi that typically exhibit alcoholic fermentation. *Saccharomyces cerevisiae* is a common yeast used in baking breads and producing alcoholic beverages such as beer and wine.

Wine production is based on the premise of yeasts fermenting the sugars found in the juices, typically grape juice, in which the yeasts are growing. In this experiment the yeasts will be allowed to ferment for only a short period of time; however, it is sufficient to demonstrate the changes that fermentation produces.

1. Obtain a 250-ml Erlenmeyer flask.
2. Pour 200 ml of grape juice into the flask.

3. Using pH test strips or a pH meter, record the initial pH of the juice below.

4. Add 5 ml of a broth culture of *S. cerevisiae* (usually grown in Sabouraud broth) to the grape juice.
5. Seal the flask by stretching the mouth of a balloon over the opening of the flask.
6. Incubate at room temperature for 1 week.
7. What changes do you observe in the appearance of the flask?

8. Remove the balloon. Using a pH meter or test strips, record the final pH below.

9. How can you explain the pH results?

COMPARE AND CONTRAST

1. How is yeast fermentation similar to bacterial fermentation? List at least two similarities.

2. What was the purpose of the balloon in the wine production experiment? Why was it not necessary in the production of yogurt?

▪ Activity 9.4: Cellular Respiration

Plants rely primarily on cellular respiration to produce the amounts of ATP needed to sustain all their activities. Remember, plants give off carbon dioxide as a byproduct of cellular respiration; however, in photosynthesis, carbon dioxide is a reactant and is used to make carbohydrates and other organic products. Carbon dioxide can dissolve in water, forming carbonic acid in the process. Because pH indicators can be used to monitor the pH of a liquid, they can be indirectly used to monitor the process of cellular respiration.

The aquatic plant *Elodea* is commonly grown in aquariums. It undergoes photosynthesis and cellular respiration. Because both processes involve carbon dioxide, the experiment must include controls to judge the impact of either process on the pH of the water. The pH indicator used here is **bromothymol blue** (BTB). At a neutral pH BTB is green. In basic conditions BTB turns blue. At acidic pH levels BTB turns yellow.

1. Obtain four test tubes.
2. Fill each test tube three-fourths full with spring water or aged tap water.
3. Add enough 0.04% BTB solution to the water until it turns green. Usually 2 to 3 ml is sufficient.
4. Add one sprig of *Elodea* to two of the tubes. Leave the other tubes with water and BTB solution only (controls).
5. Place one *Elodea*-containing tube and one control tube under direct light.
6. Place the other *Elodea*-containing tube and control tube in the dark. Wrap in foil if necessary to prevent light from entering.
7. Incubate for 1 hour at room temperature.
8. Record the appearance of water in each tube below.

   a. *Elodea* in light

   ....

   b. Control in light

   ....

   c. *Elodea* in dark

   ....

   d. Control in dark

   ....

9. Why include the controls?

   ....

   ....

   ....

10. What do the two tubes containing *Elodea* indicate concerning pH and photosynthesis versus respiration?

    ....

    ....

    ....

COMPARE AND CONTRAST

1. Compare and contrast the pH indicators used, phenol red and BTB, with respect to color changes.

2. Contrast the results seen in the two *Elodea*-containing tubes.

3. How can pH indicators be used to test for both fermentation and respiration?

4. Why is respiration not considered to be a process that produces acids, whereas fermentation does produce acids?

Study Guide

- Be able to define the terms in bold.
- Be able to describe the importance of cellular respiration and fermentation.

- Be able to identify the key steps in cellular respiration along with their intracellular locations.
- Be able to explain the importance of glycolysis, the Krebs cycle, and the electron transport system in cellular respiration.
- Be able to explain why oxygen is needed in aerobic cellular respiration.
- Be able to describe the uses of pH indicators in monitoring fermentation and cellular respiration.

Conclusions

1. What are the major products of cellular respiration?

2. What happens to the carbons and electrons that are harvested from glucose during cellular respiration?

3. Based on your observations, list three differences between bacterial and yeast fermentation.

4. Explain why pH indicators can be used to monitor cellular respiration, fermentation, and photosynthesis.

5. List three ways in which the processes of fermentation and cellular respiration are similar.

6. Which bacterial species fermented the most different types of sugars? Which species fermented the least number of sugars?

7. How does yeast fermentation affect the pH of the grape juice/wine? Given what you know about fermentation, why is this not surprising?

8. What pH indicator was used to assess cellular respiration? What is another way you could perform the experiment and avoid using that pH indicator?

9. What color indicated a positive result for sugar fermentation in the fermentation tubes? How does that compare with the positive color for the cellular respiration experiment?

10. What are two of the advantages of cellular respiration over fermentation? What are two advantages of fermentation over cellular respiration?

Chapter 10

Water Pollution

Laboratory Activities

Activity 10.1: Effects of Water Pollution on Algae

Activity 10.2: Effects of Water Pollution on Microbial Community Structure

Goals

Following this exercise students should be able to

- Describe the effects of mineral pollutants on algal growth.

- Understand the effects of mineral pollutants on the growth of microbial communities.

- Explain the role of mineral pollutants in eutrophication of water.

Introduction

Water pollution is a growing area of concern for conservation biologists and ecologists. If you recall the water cycle, runoff from rain, snow melt, and other forms of precipitation return to streams and rivers and eventually make their way to the oceans. As the water passes over the land, it picks up a variety of organic and inorganic chemicals. **Pesticides** from agricultural and domestic use along with **ammonia** and **urea** from livestock wastes are among some of the organic materials that enter water systems in this way. The inorganic chemicals include **phosphorus** and **nitrogen** from agricultural and residential fertilizers.

Concerns vary about inorganic and organic chemicals that enter the water cycle. The primary issues concerning organic materials such as pesticides are the unintended effects that may occur as nontarget species become exposed to those chemicals. With inorganic pollutants the concerns are quite different. Many of these inorganic minerals are essential nutrients for plants and algae. Thus, when these nutrients are available in excess amounts normally found in nature, these producers, especially algae, grow at increased rates. This results in what is typically referred to as an **algal bloom**. Many potential problems are associated with algal blooms. These range from the production of toxic substances to a decrease in dissolved oxygen levels that inhibit the survival of most fishes. Dissolved oxygen levels drop because of the overgrowth of algae, which produces an increase in the amount of organic material in the body of water. As that organic material decomposes oxygen levels drop, thus producing **hypoxic** conditions. The process by which excessive nutrients stimulate increased growth of plants and algae is called **eutrophication**. Eutrophication happens gradually over time; however, human activity and water pollution can greatly increase the speed at which eutrophication develops.

Activities

■ Activity 10.1: Effects of Water Pollution on Algae

Plants and algae require many essential nutrients. Those found in quantities that limit the natural growth rate of the plant are referred to as **limiting factors**. Whenever these normally limiting nutrients are found in high amounts, plants respond with increased growth rates until some other required factor becomes limiting.

Like plants, algae respond to increases in limiting nutrients by increasing growth rates. Unlike plants, increased growth in algae does not simply refer to an increase in the size of the organism. Instead, growth rate is measured as an increase in the number of cells present. Nutrients such as phosphorus and nitrogen are typically limiting in many natural algal habitats.

1. Obtain a pure culture of a unicellular alga such as *Euglena* or *Chlamydomonas*.
2. Use this pure culture to start six subcultures as described. (Note: This can be done as a class activity with each student group making separate observations.)
   a. Spring water or aged tap water (control)
   b. Spring water enriched with dissolved fertilizer (approximately 2.5%w/v)
   c. 2.5% ammonium nitrate in spring water
   d. 2.5% urea in spring water in spring water
   e. 2.5% ammonium phosphate in spring water
   f. 2.5% sodium phosphate in spring water (or 1× phosphate buffered saline, if available)
3. Incubate each subculture at room temperature in a light-filled location for 1 to 2 weeks.
4. Make a wet mount of the control culture and observe the algae under high power. What observations can you make concerning these cells?

....

....

....

5. Observe 7 to 10 fields of view. Approximately how many cells do you see in each field of view?

....

....

....

6. In these fields of view, are the cells motile? What color are they?

....

....

....

7. Repeat steps 4 through 6 for each remaining subculture. Record your observations below.

....

....

....

....

....

....

....

....

....

....

....

....

COMPARE AND CONTRAST

1. How do the algae differ between the control and the phosphate-containing cultures?

....

....

....

....

2. How do the algae differ between the control and the nitrate-containing subculture?

....

....

....

....

3. Describe the impact of urea relative to that of ammonium nitrate and ammonium phosphate.

4. How can you account for any differences observed between the ammonium phosphate and the sodium phosphate cultures?

■ Activity 10.2: Effects of Water Pollution on Microbial Community Structure

A simple way to observe a community of organisms is to prepare a hay infusion microcosm. Hay infusions are generated by soaking dried hay or grass stalks in water for approximately 2 weeks. Any microbes present in the hay or water will begin growing and multiplying within the hay infusion. It is common to observe protozoans, algae, and bacteria in the liquid medium of the hay infusion.

1. Produce six hay infusions as described.
   a. Hay infusion with spring water or aged tap water (control)
   b. Spring water enriched with dissolved fertilizer (approximately 2.5% w/v)
   c. Hay infusion with 2.5% ammonium nitrate
   d. Hay infusion with 2.5% urea
   e. Hay infusion with 2.5% ammonium phosphate
   f. Hay infusion with 2.5% sodium phosphate
2. Make a wet mount with a sample from the control hay infusion. Observe under high power.
3. Observe 10 fields of view. Record the types and numbers of organisms viewed.

....
....
....
....
....

4. Repeat steps 2 and 3 for the remaining hay infusions. Record your results below.

....
....
....
....
....
....
....
....
....
....
....
....
....
....

COMPARE AND CONTRAST

1. Which hay infusion exhibited the most variety of organisms?

....
....
....
....

2. Which hay infusion exhibited the highest number of organisms? How does this compare with the most diverse hay infusion?

....

....

....

....

3. Compare the algal culture with the highest number of algae observed with the hay infusion exhibiting the highest number of organisms.

....

....

....

....

Study Guide

- Be able to define the terms in bold.
- Be able to describe eutrophication and its causes.
- Be able to identify the effects of increased nutrients on populations and communities.
- Be able to explain how water becomes contaminated with pollutants.

Conclusions

1. What is eutrophication?

....

....

....

2. Of the nitrogen sources tested (ammonium nitrate, ammonium phosphate, urea, and fertilizer), which had the largest impact on algal growth?

....

....

....

3. Of the nitrogen sources tested (ammonium nitrate, ammonium phosphate, urea, and fertilizer), which had the largest impact on the growth of microbial communities?

4. Which form of phosphorus had the largest impact on protozoan and algal growth?

5. Describe some potential outcomes of eutrophication.

6. What is a growth limiting factor?

7. Describe a common source of urea in the environment.

8. List three common water pollutants that can originate in residential areas.

9. Describe one concern with organic water pollutants. Can you think of an example not mentioned in this lab?

10. Describe the major concern with inorganic water pollutants.

Chapter 11

Case Studies in Mineral Nutrition

Laboratory Activities

Activity 11.1: Case Study 1

Activity 11.2: Case Study 2

Activity 11.3: Case Study 3

Activity 11.4: Case Study 4

Activity 11.5: Case Study 5

Goals

Following this exercise students should be able to

- Describe the general symptoms of mineral deficiency.

- Identify common mineral deficiencies based on symptoms.

- Understand the role of macronutrients and micronutrients in a plant's life.

Introduction

Plants need a variety of minerals to maintain their health. Some of these are needed in high quantities, the **macroessential elements,** and some in much smaller amounts, the **microessential elements.** Regardless of the quantity needed, these minerals must be present for the plant to complete its life cycle. The essential minerals for plants are listed in **Table 11-1.** Use your textbook or other resources to get more information about the impact of these minerals on plant growth.

Some essential elements are immobile. These **immobile elements,** once a part of the plant's tissues, cannot move from one area to another within the plant. They stay in their original location within the plant. **Mobile elements** are somewhat different because the plant can recycle these elements from older tissues to younger tissues. Mobility of an element is therefore helpful in diagnosing mineral deficiencies. The effects of deficiencies in mobile elements are usually seen in older tissues, whereas the effects of immobile elements are seen in young tissues.

Many **mineral deficiencies** can cause species-specific symptoms. Mineral deficiencies, however, produce similar symptoms in a variety of plants. For example, zinc, iron, and magnesium are all known to cause chlorosis, the loss of chlorophyll in leaves. **Chlorosis**

TABLE 11-1:	Essential Elements to Most Plants and Their Functions
Macronutrients	
Carbon	Organic compounds
Oxygen	Organic compounds
Hydrogen	Organic compounds
Nitrogen	Amino acids; nucleic acids; chlorophyll
Potassium	Amino acids; osmotic balance; enzyme activator; movement of guard cells and motor cells
Calcium	Controls activity of many enzymes; component of middle lamella; affects membrane properties
Phosphorus	ATP; phosphorylated sugars in metabolism; nucleic acids; phospholipids; coenzymes
Magnesium	Chlorophyll; activates many enzymes
Sulfur	CoA; some amino acids
Micronutrients	
Iron	Cytochromes; nitrogenase; chlorophyll synthesis
Chlorine	Unknown; possibly involved in photosynthetic reactions that liberate oxygen
Copper	Plastocyanin
Manganese	Chlorophyll synthesis; necessary for the activity of many enzymes
Zinc	Activates many enzymes
Molybdenum	Nitrogen reduction
Boron	Unknown
Except for boron, all these elements are essential for human beings. But unlike plants, our diet must also provide us with fluorine, iodine, cobalt, selenium, chromium, and sodium. We obtain fluorine by adding it to our drinking water (fluoridation), and iodine is obtained by adding it to salt or by eating a large amount of seafood. Our lives depend on sodium, and if we lose too much by sweating, we can quickly die. Most plants have no need for sodium at all; the exceptions are C4 plants and CAM plants, which need trace amounts.	

causes leaves to become yellow to almost white in color. The veins of the affected leaves are often unaffected and remain green. Why might that be? The differences in the symptoms of these deficiencies focus on the details of the symptoms. For instance, iron is an immobile element, whereas magnesium is a mobile element. These deficiencies appear in young leaves and older leaves, respectively. In both cases the amount of chlorosis increases with increasing distance from a vein. The chlorosis associated with zinc deficiency is often associated with the formation of brown spots where areas of the leaf tissue die.

Potassium deficiencies are most evident in changes along the margin of a leaf. The margins and leaf tips often die, turning the edges of the leaf a dark brown or black color. This color change along the margins is due to **necrosis** (death) of the cells in this area. Phosphorus deficiency causes a discoloration of the leaves; however, this is not chlorosis. Instead, the leaves experience a buildup of anthocyanin, a pigment also found in high quantity in red cabbage. This increase in anthocyanin results is a dark purple discoloration in the affected leaves.

Activities

Using what you know about mineral deficiencies in plants, identify the causes of each plant deficiency described in the cases below. Refer to your textbook if you need additional information about the deficiencies or their symptoms.

▪ Activity 11.1: Case Study 1

Leah moved to southern Florida from Michigan with her family 5 years ago. Although she has enjoyed the tropical weather and never needing a snow shovel, she still doesn't know much about caring for plants in this area. Back in Michigan she grew lovely azaleas, but she doesn't see many azaleas around Miami. She lives in a modest home with a landscaped yard, courtesy of the previous owners. Her favorite plants on the grounds are the hibiscus with their large, showy blooms all summer.

   The hibiscus plants were gorgeous with dark green, glossy foliage when she moved into the home and remained so for the first several years. This summer she notices the leaves are turning somewhat yellow. The young leaves appeared almost completely yellow, except for the area around the veins. She is distraught, believing she has managed to kill her favorite plants and asserts that she has never had a "green thumb."

1. What can you tell Leah about the cause of this discoloration of leaves?

2. Is this yellowing of young leaves reversible? If so, how should Leah treat the plants to reverse the damage?

3. This year has seen record-setting rainfall in the Miami area. Does that have any impact on your diagnosis? Why?

▪ Activity 11.2: Case Study 2

Paul is growing potato plants in his parents' backyard. For years his grandparents have been growing a large garden, and he wants to put some of the tips they have passed on to him to use in his own garden. This year Paul wants to begin by growing potatoes and a few other vegetables. Potatoes are appealing to him because he has heard they are easy to grow and do well in the central Texas area. Paul is new to gardening and is trying to follow his grandparents' instructions carefully.

As Paul looks forward to harvesting his potatoes, he begins to notice some changes in the 2-month-old crop. The edges of the leaves have begun to turn dark and look dried out, almost like they've been burned. At first, Paul doesn't think much of this difference, but after a couple of weeks all the plants look the same way and the majority of the leaves are rimmed with a black edge. Paul is afraid his potatoes are dying.

1. Paul doesn't know what to do to salvage his plants and doesn't want to admit failure by calling his grandparents for help. So, he does some Internet research and determines a likely cause of the symptoms. What is that cause?

2. What does Paul need to give his plants to help them recover?

3. If Paul has correctly identified the deficiency, what do you expect the plants to look like a week after treatment?

■ Activity 11.3: Case Study 3

Xin has never had any experience with plants but wants to grow tomatoes and peppers on his patio to make truly homemade sauces. He has found many websites with growing instructions as well as recipes, and he is anxious to get started. He starts by selecting several varieties of peppers from a local nursery. He prices potting soil but decides it is just too expensive, so he fills his containers with soil from a vacant lot across the street from his apartment complex. He carefully removes the larger stones from the soil and fills his containers. He notices the soil is light tan in color but doesn't think too much about it.

Xin gets all his peppers planted and watered. He checks the peppers everyday and follows the watering instructions that came on the tags. He wants to be sure he can grow plants before he buys the tomatoes. After 3 weeks of caring for his peppers Xin notices the older leaves are starting to look rather strange. The veins are still the normal green color, but the rest of the leaf has turned yellow. Xin is concerned about his pepper plants. They are still growing and still producing new leaves, but these older leaves just look really strange.

1. What can you tell Xin about the most likely cause of these symptoms?

2. What term is used to refer to the yellowing of leaves?

....

....

....

3. How should Xin treat his plants to ensure their recovery?

....

....

....

▧ Activity 11.4: Case Study 4

Maria has just bought her first house on the outskirts of the greater Chicago area. Her home is situated in a new development that was built in an area that had once been a large cornfield. She is ecstatic about being truly independent with a good job and a new home. She has wanted to try growing her own fruits and vegetables for several years and now sees the perfect opportunity. After all, if this was agricultural land a couple of years ago, it must make for wonderful garden soil.

Maria is growing beans in her backyard garden. Recently, she noticed that many of the bean plants had leaves that were beginning to turn pale green, even yellow toward the edges. As she continued to watch the plants, making sure they received sufficient water, she noticed that some of these pale areas of the leaves started turning brown. It looked as if the leaves had brown freckles, usually clustered in areas that had lost the typical dark green coloration.

1. What could be causing this type of discoloration?

....

....

....

2. What would you recommend Maria do to counteract this problem?

....

....

....

▧ Activity 11.5: Case Study 5

John is a college student in the Phoenix area and is growing a container garden for the first time this year. He has heard about the advantages of organic gardening and is applying these techniques to his gardening. Following the guidelines for organic gardening, John has not applied any chemical pesticides or fertilizers to his garden. He is growing tomatoes, peppers, and beans in a small garden in his backyard.

All the plants in John's garden grew quite well for the first 2 or 3 weeks. As time passed John started to notice some discoloration in the tomato plants. The tomato plants have begun to grow more slowly and do not look as healthy as they did when John purchased them from the local market. The plants have also begun to develop dark purple–colored areas on their leaves. John is upset that his tomatoes are not performing well and is getting frustrated with gardening.

1. What mineral(s) is missing from John's tomatoes?

2. How can that mineral be supplied to the plants?

3. Can John improve the health of the plants and still garden organically? If so, how?

COMPARE AND CONTRAST

1. Using the resources available to you, explain why the symptoms of iron and magnesium deficiencies are similar, other than in location.

2. Deficiencies of phosphorus, potassium, and iron result in darkening of the leaf. Compare the symptoms seen in each case.

Study Guide

- Be able to define the terms in bold.
- Be able to identify the symptoms of each mineral deficiency.
- Be able to describe mobile and immobile elements and know examples of each.
- Be able to answer each of the questions asked in the lab exercise.

Conclusions

1. What makes an element "essential" to a plant?

2. How do macroessential elements differ from microessential elements with respect to the plant's needs?

3. Why is chlorosis damaging to the plant?

4. What is the advantage to the plant of being able to move certain elements to different locations within the plant?

5. From what source do plants procure these minerals that are being moved from one location to another within the plant?

6. How are biogeochemical cycles important in providing minerals to plants? (Note: you may need to refer to your textbook for additional information.)

7. What human activities increase the likelihood of mineral deficiencies developing in plants?

8. Most fertilizers contain a combination of nitrogen and potassium. Why do fertilizers not need to incorporate carbon?

9. Describe how symptoms of deficiency can help a person identify the types of minerals that may be in insufficient supply for the plant.

Chapter 12

Tissue Culture

Laboratory Activity

Activity 12.1: Growing Plant Cells in Tissue Culture

Goals

Following this exercise students should be able to

- Describe the importance of tissue culture to plant cultivation.
- Explain the mechanism by which plant cells can be grown in tissue culture.
- Understand how the genetic composition of plant cells relates to plant propagation by tissue culture.

Introduction

Seed plants can be propagated through a number of means. The traditional means of plant propagation includes harvesting and planting seeds. This method has been used throughout human history to domesticate crop plants and to move plants from one geographical region to another. Seeds are produced as a direct result of **sexual reproduction** and contain dormant embryos that will grow and produce the next generation of plants. The benefit of sexual reproduction is the increase in **genetic variation**, which produces plants that are slightly different from one another. This genetic variation is essential for the species to survive changing climatic conditions.

Most seed plants can also be propagated asexually. Some plants naturally produce stolons, or runners, that spread the plant horizontally through its habitat. This is a simple type of **asexual reproduction**. With human involvement asexual plant propagation can take many other forms. Allowing cuttings to develop roots is a widely used, simple means of asexual plant propagation. The benefit of asexual plant reproduction is that all the plants generated in this manner are genetically identical; thus, the grower knows exactly what traits the plants will possess.

Cuttings are able to grow into intact plants because the parenchyma cells in the stem or leaf are capable of cell division. Because each cell contains all the genetic information needed to produce an intact plant, a condition called **totipotency**, the cells produced by this division can become any other type of cell the plant needs. These genes will be passed from the parental cells to their offspring during cell division, so one cell capable of dividing is sufficient to produce all the cells necessary for the plant to survive. As the cells increase in a plant, they begin differentiating and taking on specific functions.

This developmental process begins in nature with the formation and development of an embryo from a zygote.

Another type of asexual plant propagation is done through **tissue culture**. Tissue culture involves taking a sample of cells from an original organism and growing those cells in liquid media in a laboratory setting. This technique is a relatively new means by which plants can be propagated. The major advantage of tissue culture for plant propagation is the large numbers of identical plants that can be generated in a relatively short period of time.

Tissue culture can be used for a variety of goals. In its simplest form it is a way to harvest numerous cells from one original individual and allow each cell to grow into a fully formed plant. With advances in genetics, plant cells can now be genetically modified while grown in tissue culture. The cells that were successfully modified then grow into genetically modified plants (**Figure 12-1**).

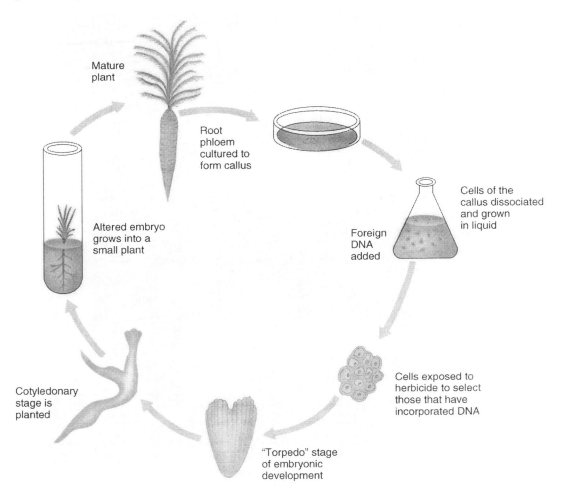

FIGURE 12-1   A tissue culture method using recombinant DNA techniques to genetically alter a plant. (© Jones & Bartlett, LLC)

Activity

▪ Activity 12.1: Growing Plant Cells in Tissue Culture

Many plant tissues can be used as starter cells for tissue culture. Root vegetables, such as carrots or radishes, are used in this exercise. Maintaining a sterile environment is the key to successful tissue culture. If available, all steps should be completed in a biological safety hood. If a hood is not available, tissue culture may be completed through bench-top work; however, significant bacterial and fungal contaminants can be expected.

The equipment used in this experiment needs to be sterilized between each use. Forceps and scalpels can be sterilized by dipping them in 70% alcohol and passing them through the flame of a Bunsen burner. This is called "flaming" the equipment.

1. Obtain a small section of carrot taproot. Taproots that are intact and undamaged have sterile interiors and work best for this procedure.

2. Using a sterile knife or scalpel, cut a 3- to 6-cm section from the middle of the taproot. Discard both ends of the root.

3. With a sterile scalpel, remove the epidermis from the root section, along with any visible blemishes or damaged areas.

4. Sterilize the surface of the carrot tissue by soaking the tissue in 10% sodium hypochlorite (bleach) solution in sterile water for 5 minutes.

5. Pour off the bleach solution into a waste container.

6. Wash the root section three times with sterile water.

7. Transfer the root section to a fresh, sterile Petri dish.

8. Cut a transverse section approximately 2 to 3 mm thick. Transfer the section to a sterile Petri dish.

9. Cut the transverse section into smaller sections by cutting across the cambium. To do this, lay the transverse section flat. Make two cuts that divide the width of the section into thirds. Discard the two sections on the outer edges. Turn the remaining section lengthwise. Make two more cuts to divide that section into thirds. Discard the two sections on the outer edges. You should now have a square section from the middle of the carrot root.

10. Subdivide the square, central section of the taproot into three equal-sized strips. The two outer strips are ready to use. Cut the middle strip in half midway through the length of the strip.

11. Transfer each of these small sections to separate, sterile Petri dishes containing a complex, sterile medium such as Murashige & Skoog.

12. Seal each plate with Parafilm or tape to ensure the lids remain in place.

13. Incubate the plates at room temperature in dark conditions.

14. Check the plates weekly and monitor the growth of the tissue explants. Use the space below to keep records of appearance. If an inverted microscope is available, observing the cells under low and high power will be helpful.

    a. Week 1:

    b. Week 2:

    c. Week 3:

    d. Week 4:

    e. Week 5:

Within 4 to 5 weeks you should observe a callus being produced. The callus appears as a large mass of cells, approximately 2 to 5 cm in diameter, growing from the original root section. This callus can be maintained in tissue culture by subculturing (removing a small portion of the callus to a sterile Petri dish with fresh, sterile media). By subculturing the callus you can generate a large number of plants from the initial tissue sample, all of which are genetically identical.

COMPARE AND CONTRAST

1. How did the appearance of the cells change between weeks 1 and 2?

2. How did the appearance of the cells change between weeks 1 and 5?

3. What would you expect to happen if you left the callus in the plate for an additional 5 weeks?

Study Guide

- Be able to define the terms in bold.
- Be able to explain the reasons tissue culturing techniques are useful in botany.
- Be able to describe the process of growing cells in tissue culture.
- Be able to describe the changes that occurred as the plants grew in the tissue culture medium.

Conclusions

1. What does totipotency mean?

2. Why are sterile tools so important for starting a tissue culture?

3. Based on your results, when did the callus begin to form?

4. Why did the initial samples for tissue culture come from the center of the root instead of at the edge?

5. Why is sealing the edge of the Petri dish important?

6. What tissue culture medium was used?

7. Describe one major benefit of sexual reproduction.

8. Describe one major benefit of asexual reproduction.

....

....

....

9. Would you expect an animal tissue sample, such as muscle, to be able to produce an entire organism through tissue culture techniques? Why or why not?

....

....

....

10. Explain the process of subculturing.

....

....

....

Chapter 13

© Cristi Matei/ShutterStock, Inc.

Genetics, Inheritance, and Natural Selection

Laboratory Activities

Activity 13.1: Predicting the Outcomes of Crosses

Activity 13.2: Population Genetics

Goals

Following this exercise students should be able to

- Describe flow of information in a cell.
- Understand the roles of DNA and proteins in the cell.
- Use a Punnett square to predict the outcomes of a genetic cross.
- Describe the influence of natural selection on populations.

Introduction

Genes are sequences of nucleotides in DNA that code for the production of a protein or RNA molecule. These are the information-containing units that are passed from one generation to the next. Genes direct the actions of a cell by directing the production of **proteins**. Proteins serve a variety of purposes in cells, including structural support, movement, and enzymatic activity. They create the various anatomical and physiological traits of the organism.

The genes contained within an organism's chromosomes make up that organism's **genotype**. The traits produced by the expression of those genes form the organism's **phenotype**. Although the genotype of an individual determines its phenotype, one cannot always determine genotype by observing an organism.

Heterozygous individuals contain two different versions of the same gene. These alternate gene versions are called alleles. When an individual contains two contrasting alleles, one may mask the presence of the other. The one whose presence is masked is said to be **recessive**, whereas the other is **dominant**. **Homozygous** individuals contain two identical alleles. If the individual is homozygous dominant, the dominant phenotype is present. In contrast, if the individual is homozygous recessive, the recessive phenotype is apparent.

Punnett squares are tools used to predict the outcomes of genetic crosses. In a one-trait cross the possible gametes from two individuals of known genotypes are used to predict the offspring. The simplest type of Punnett square is used to predict the offspring of a one-trait cross. To complete a Punnett square, you have to know all the potential gametes.

For instance, if you have a homozygous tall plant (TT) and a homozygous short plant (tt), each can produce one type of gamete. The homozygous dominant plant will produce gametes carrying the dominant allele (T), whereas the homozygous recessive plant will produce gametes carrying the recessive allele (t). The process is a bit more complicated when dealing with heterozygotes that carry one dominant and one recessive allele (Tt). In this case, 50% of the gametes carry the dominant allele and 50% carry the recessive allele. Once you have determined the gametes, you are ready to set up your Punnett square (see example in **Figure 13-1**). The possible gametes from one parent are placed along the top of the square, and the possible gametes from the other parent are placed along the left side of the square. The boxes represent the possible offspring. The gametes at the top and along the side of the square are carried into the corresponding boxes. The gametes at the top of the square are added to each box below and the gametes from the left are added to each box in the same row. In the end each box should contain two gametes.

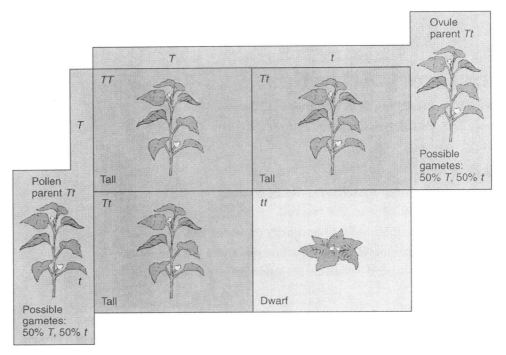

FIGURE 13-1   In setting up a Punnett square for a selfing of Tt plants, first establish the genotypes of the two parents. Then determine what types of gametes are produced and in what proportions and fill in the squares with the genotypes. From the genotypes the phenotypes in each square can be determined. (© Jones & Bartlett, LLC)

Remember, the boxes in the Punnett square represent all the possible offspring. Collectively, these four boxes represent 100% of the offspring. By comparing the number of boxes with the same genotypes represented, you can calculate the likelihood of one offspring having a particular genotype. For instance, if half of the boxes have heterozygous genotypes, there is a 50% chance the offspring will be a heterozygote.

Punnett squares can also be used to predict offspring in more complicated crosses. Two-trait crosses involve tracking the inheritance of two genes. The process is the same; however, the Punnett square has 16 boxes instead of four. To find the possible gametes in a two-trait cross, you have to account for all the possible alleles of a combination of two genes. For example, a plant that is heterozygous for height (Tt) and for yellow seeds (Yy) can produce four types of gametes: TY, Ty, tY, and ty.

Inheritance of genes is important because it directs the development of the individual. On a population level, genes ensure that each member of the same species have similar traits. At the same time, each individual contains a unique complement of genes, so each individual will have a slightly different phenotype. This natural variation results in an assortment of individuals that have different chances for surviving and reproducing. Those that are successful in producing more offspring essentially donate more copies of their alleles to the subsequent generation than do those that do not produce as much offspring. **Natural selection** is the process that determines which of these traits is beneficial and which is harmful.

The process works on all traits an organism possesses that impact survival and reproduction. The result of natural selection is adaptation. **Adaptations** are traits that cause a population to be better suited for its environment. This is the core concept of biological evolution. It is important to understand that populations evolve through the process of natural selection. Individual organisms do not evolve in the biological sense of the term. Natural selection impacts individual survival rates and reproductive success, but the result is the evolution within the entire population as beneficial traits become more prevalent.

The genetics of a population can be influenced by many factors. Genetic variation is increased by continual crossing-over during meiosis and by the random fertilization of gametes. Other factors have an impact as well. **Migration** of organisms between populations provides opportunity for new allelic variants to enter a population. Migration can also result in the loss of alleles from a population.

Alleles can also be lost from a population through **genetic drift**, random events that result in the loss of individuals from a population. One type of genetic drift is the **founder effect** where a small group of individuals leaves a parent population to establish a new population in a discontinuous environment. **Bottlenecks** can occur when a large percentage of a population is killed by a chance event, such as an earthquake. This leaves a remnant of the original population with some subset of the original alleles. Bottlenecks are associated with an overall loss of genetic variation from within a population.

Activities

▪ Activity 13.1: Predicting the Outcomes of Crosses

We use beads here to randomly generate genotypes to predict offspring of genetic crosses. Red beads represent dominant alleles, whereas white beads represent recessive alleles.

1. Put 10 red beads and 10 white beads into a beaker. Mix them well.
2. Close your eyes and pick two beads out of the beaker.
3. Record the genotype of the first organism below:
   a. Organism 1

   ....

4. Repeat the process three more times and record the genotypes below:
   a. Organism 2

   ....

   b. Organism 3

   ....

   c. Organism 4

   ....

5. Complete a Punnett square below for a cross between organism 1 and organism 2.
    a. Punnett square for offspring 1:

6. Complete a Punnett square below for a cross between organism 3 and organism 4.
    a. Punnett square for offspring 2:

7. Choose one genotype from the Punnett square for offspring 1. Record that genotype here.

8. Choose one genotype from the Punnett square for offspring 2. Record that genotype here.

9. Use a Punnett square to predict the offspring of offspring 1 and offspring 2. Show your work below.

10. Which trait is the most common in your resulting offspring?

11. Compare your results with the rest of the class. Overall, which trait is more common among the offspring shown in step 9?

COMPARE AND CONTRAST

1. What is the difference between a dominant allele and a recessive allele?

....

....

....

....

....

2. Assume the offspring in the Punnett square shown above in step 9 represents the genotypic ratios of the entire population. Is the dominant trait always the most common in a population? Why or why not?

....

....

....

....

....

▪ Activity 13.2: Population Genetics

We use beads here to simulate the individuals within a population. Red beads represent dominant traits, and blue beads represent recessive traits. Each bead represents one plant. The beads will be placed on a black-and-white patterned background (paper or fabric) approximately 1 square yard in size. Each color in the background represents different soil types and microenvironments.

1. Obtain 15 red beads and 15 blue beads. Mix them together in a beaker.

2. Distribute the beads randomly on the background pattern. Beads that land on white background represent seeds that germinate in rocky soil capable of holding little moisture. Beads that land on dark background represent those that germinate in fertile soil capable of retaining a moderate amount of water.

3. Assume that drought conditions prevail during the next summer. Plants in the rocky soils die. Remove any beads that landed with a majority of the bead in a white area.

4. The others survive the drought and reproduce. For simplicity, assume each plant produces one offspring that survives during the next growing season. In effect, you will double the number of beads remaining on the fabric. Do not remove any additional beads. Count the number of blue and red beads remaining.

   a. How many red beads remained after the drought (before reproduction)?

....

    b. How many blue beads remained after the drought (before reproduction)?

....

    c. Did the plants survive based on a genetic factor or based on a random event?

....

    d. How can you best describe what happened in this scenario in terms of population genetics?

....

5. Add equal numbers of blue and red beads to the beaker. Mix and distribute the new beads randomly over the same background.

6. The next growing season has optimal moisture conditions. Plants in all soil type survive. During the following winter temperatures drop to record lows of −18°F and remained below 5°F for 3 consecutive days. One small area of the environment was near a natural hot spring and did not experience the extreme cold temperature for a prolonged period. Identify this area by randomly tossing a 2-inch square piece of paper onto the background. Any beads touching this piece of paper will automatically survive.

    a. How many red beads touch the paper?

....

    b. How many blue beads touch the paper?

....

7. The blue beads represent plants that are capable of producing a compound that prevents ice crystal formation within the plant—a natural "antifreeze" compound. This is a recessive trait. Red beads represent plants with no protection from freezing and will be damaged and/or killed by ice crystal formation. Remove all red beads (other than those touching the paper in step 6) because they did not survive the winter.

    a. How many blue beads remain in the area not touching the paper?

....

8. Again assume each plant that survived the winter successfully produced one offspring in the spring. Count the number of beads remaining (totals from steps 6 and 7).

    a. How many red beads remain?

....

    b. How many blue beads remain?

....

9. Distribute an equal number of new beads randomly on the background.

10. Assume the following growing season has optimal rainfall followed by a mild winter (no temperatures below 10°F). All the plants survive.

    a. How many red beads are present?

....

    b. How many blue beads are present?

....

11. Assuming that each plant experiences replacement reproduction (one offspring per plant), how many beads will be present after the following reproductive season? (You do not need to add more beads to the background at this time.)

    a. How many red beads are present after reproduction?

    b. How many blue beads are present after reproduction?

12. Is this trend due to a genetic difference between the plants or to random events?

13. How can you best describe this event in terms of population genetics?

14. Assume the following population from step 10 experiences a growing season with moderate amounts of rainfall sufficient for all the plants to survive; however, a wildfire burns the top right quadrant of the area, killing all the plants in that region.

    a. How many red beads remain after the fire?

    b. How many blue beads remain after the fire?

    c. Did the plants die based on a genetic factor or on a random event?

    d. How can you best describe what happened in this scenario in terms of population genetics?

COMPARE AND CONTRAST

1. How do you expect the frequency of the recessive (exhibited by the blue beads) trait to change if the environment continues to experience severe winters?

2. How do you expect the frequency of the recessive (exhibited by the blue beads) trait to change if the environment continues to experience warm winters?

Study Guide

- Be able to define the terms in bold.
- Be able to describe the genetic factors that can influence a population.
- Be able to explain the link between genetic variation, natural selection, and evolution.
- Be able to use a Punnett square to predict the offspring of one-trait and two-trait crosses.

Conclusions

1. Describe the process of natural selection.

2. Compare genotype and phenotype.

3. If you know the phenotype of an organism, does that mean you know its genotype? Why or why not?

4. List and describe three events that increase genetic diversity in a population.

5. What happens in genetic drift?

6. Use a Punnett square to predict the outcome of a cross between a heterozygous plant with yellow seeds and a homozygous recessive plant with green seeds.

7. Use a Punnett square to predict the outcome of a cross between a heterozygous white-flowered plant and a homozygous red-flowered plant.

8. List the possible alleles that can be produced by a plant heterozygous for seed color (yellow = dominant, green = recessive) and seed texture (smooth = dominant, wrinkled = recessive).

9. Describe the flow of information from genes to phenotype.

10. Under what circumstances might a dominant allele not become common in a population?

Chapter 14

Classification and Systematics

Laboratory Activities

Activity 14.1: Artificial Classification

Activity 14.2: Natural Classification

Goals

Following this exercise students should be able to

- Design and use a dichotomous key.

- Define the role of classification in botany.

- Compare artificial and natural classification systems.

- List the hierarchical organization of taxa in order of increasing specificity.

Introduction

Classification refers to the grouping and naming of organisms. This has been done for centuries as humans have attempted to make sense of the natural world. **Systematics** is a newer field of science that seeks to group organisms based solely on relatedness. Modern classification of organisms is based on a systematic approach. Relatedness is determined by comparative anatomy, comparative physiology, biochemical analyses, and other tests. Once scientists know how an organism is related to other similar organisms, it is classified and named based on that information.

Organisms can be classified using a variety of systems. The two most commonly used systems for classifying plants are artificial and natural classification systems. **Artificial classification** systems group organisms based on similar features, not on relatedness. **Natural classification** systems seek to classify organisms based on relatedness. It includes the assignment of unique names to each organism.

A **scientific name** consists of the **genus** name and the **specific epithet**. A genus (plural, genera) is a group of related species. **Species** are classified into genera using a natural classification system. The specific epithet allows one to identify one species from another within the same genera. This two-word combination, or **binomial**, is unique for every organism. Related genera are grouped into **families**. Families are grouped into **orders**, orders into **classes**, classes into **phyla** (singular phylum), phyla into **kingdoms**, and kingdoms into **domains**. These **taxa** represent the hierarchy of classification. A taxon is defined as a group of related organisms. It can be at any rank in the hierarchy of classification.

Artificial classification systems are more commonly used as methods of identifying known species. Some examples of such systems are guides to roadside wildflowers and books for identifying backyard birds. They often group organisms by similar physical features that do not necessarily reflect relatedness. For example, all white-flowering plants may be grouped together and kept separate from the red-flowering species.

Activities

▪ Activity 14.1: Artificial Classification

Artificial classification systems are typically used for identification of plants or animals. One type of artificial classification system is a **dichotomous key**. Dichotomous keys are generally written as a series of paired statements. Each statement in the pair provides contrasting descriptions of plant characteristics. Thus, a dichotomous key is used as a process of elimination based on visible traits that is used to identify the species of organism being studied. Each statement must provide an answer (i.e., plant name) or directions for the reader to proceed to another set of statements.

Many dichotomous keys are available for identifying plant species. Some are based on woody twig anatomy, some are based on leaf anatomy, and others are based on flower structure or fruits. Complex dichotomous keys often use a variety of anatomical characteristics. Dichotomous keys for plant identification are based on visible characteristics, not metabolic properties. They can be taken into the field and used to identify a variety of plants. Often, dichotomous keys are limited to the plants commonly found in a particular geographical region. This prevents the keys from becoming too cumbersome.

Below is a partial dichotomous key for identifying types of fruit to use an example. The key you develop will be different; however, the format will be the same.

1. Fruit is simple. . . . . #2
1. Fruit is complex (multiple, aggregate, or accessory) . . . . . . #5

2. Fruit is fleshy. . . . . #3
2. Fruit is dry. . . . . #10

3. Fruit has a hard, stony pit surrounding the seed. . . . . Drupe
3. Fruit has seeds suspended in flesh. . . . . #4

4. Fruit contains multiple seeds and has a thin skin. . . . . Berry
4. Fruit contains multiple seeds and has a thick rind. . . . . Hesperidium

Your task is to design your own dichotomous key. Remember, format is important. Statements should be written in contrasting pairs and should direct the reader to an answer or to the next relevant set of questions.

1. Obtain seven preserved leaf specimens, or photographs of leaves if preserved material is not available.
2. The easiest way to start building a dichotomous key is to separate your specimens into two sets based on some characteristic of their leaves that differs between the two groups. What characteristic are you using to form these initial groups?

3. Using the two groups you've just created and the key difference between them, formulate your first two statements. Write them below. One statement should describe each group and they should be contrasting with reference to the key leaf characteristic listed in step 2.

....

....

....

4. Based on leaf structure alone, build a dichotomous key. Do not use the presence, absence, or type of fruit present, stem characteristics, or any other physical trait. Refer to the lab exercise on leaves or your textbook for descriptive terms used with leaves.

....

....

....

....

....

....

....

....

....

....

....

....

....

....

....

....

....

■ Activity 14.2: Natural Classification

Natural classification systems can be used to build a phylogeny, the evolutionary history of a group of organisms. **Phylogenies** are depicted as branched diagrams, called **phylogenetic trees**, with the common ancestor forming the root, or base, of that diagram. Along the branches of each phylogenetic tree, the characters that set one group apart from another are listed. This allows the reader to quickly understand why one lineage is grouped separately from the other.

Phylogenetic trees are constructed using a principle called **parsimony**. Simply stated, parsimony means the simplest answer is most likely correct. In an evolutionary sense, that means a specific trait, such as bilaterally symmetrical flowers, likely evolved once in the lineage of a particular group of plants. Thus, you should not see a trait appearing, disappearing, and reappearing within the phylogeny of one group of organisms. Each trait most likely evolved only once, although some branches of this lineage may have lost the trait.

The goal of natural classification systems is to group organisms into **monophyletic** groups that contain one ancestor and all of its descendents. Sometimes, only some of an ancestor's descendents are included within one group. This is a **paraphyletic** grouping. **Polyphyletic** groupings are the least desirable. In these groups, organisms are grouped together that do not share a recent common ancestor. Organisms in paraphyletic groups therefore may not be closely related at all.

Each branch of a phylogenetic tree ends with the name of a plant species, and notations are made along the branches depicting the key characteristics of the species along that branch. Remember, phylogenies are based on relatedness. The root of the phylogenetic tree represents the ancestor to everything else on that tree.

To design a phylogenetic tree you must first determine which features or characteristics are ancestral. Typically, **ancestral traits** are found in the majority of the related species. Those traits may be modified in some groups, but they serve as characteristics that unite all these related species. These are characteristics that would have been present in the most recent common ancestor of all these species, represented by the root of the diagram. The most recent common ancestor may not be identified; however, it is still represented by the root of the tree. Characteristics that are shared by some, but not all, of the descendent species are called **derived traits**. Derived traits are useful for determining subgroups within a phylogeny.

1. Obtain a selection of hardware (nails, screws, bolts, staples, etc.). You should have approximately 10 to 15 pieces. Each piece of hardware represents one species.

2. Spread out the hardware so you can see all the pieces at once. What are the ancestral characteristics? Name at least one.

3. Which "species" of hardware most closely match the ancestral condition?

4. Use these "species" of hardware to construct your own phylogenetic tree. Pay attention to the format. Make note of the key characteristics on each branch of the diagram.

COMPARE AND CONTRAST

1. How are dichotomous keys similar to phylogenetic trees? How are they different?

2. Describe the benefits of artificial and natural classification systems. Which is more useful to you? Why?

3. Does your phylogenetic tree exhibit maximum parsimony? Why or why not?

Study Guide

- Be able to define the terms in bold.
- Be able to correctly use the terms polyphyletic, monophyletic, and paraphyletic.
- Be able to read a dichotomous key and a phylogenetic tree.
- Understand the role of artificial and natural classification systems in botany.

Conclusions

1. Describe the basis for formulating an artificial classification of organisms.

2. Describe the basis for formulating a natural classification of organisms.

3. What is parsimony?

4. How is parsimony useful in developing phylogenies?

5. Which is the preferred grouping: monophyletic, polyphyletic, or paraphyletic? Why?

6. Kingdom Protista is polyphyletic. Based on your understanding of phylogeny, what does this mean?

7. List the hierarchical taxa used for classifying organisms from the broadest (most inclusive) to the most specific.

8. How do derived and ancestral traits differ?

9. How are dichotomous keys useful in botanical studies?

    ....

    ....

    ....

10. Currently, all plants are classified within one kingdom because they share the same basic traits. Is this a monophyletic taxon or a polyphyletic taxon? How can you tell?

    ....

    ....

    ....

Chapter 15

Algae

Laboratory Activities

Activity 15.1: Division Chlorophyta—The Green Algae

Activity 15.2: Division Rhodophyta—The Red Algae

Activity 15.3: Division Phaeophyta—The Brown Algae

Activity 15.4: Division Chrysophyta—Diatoms and Their Relatives

Activity 15.5: Division Pyrrhophyta—The Dinoflagellates

Activity 15.6: Division Euglenophyta—The Euglenoids

Activity 15.7: Division Cyanobacteria—The Blue-Green Algae

Goals

Following this exercise students should be able to

- Describe the general characteristics of algae.

- Recognize the species of algae studied.

- Understand algal classification.

Introduction

Algae are eukaryotic, aquatic, photosynthetic organisms classified in Kingdom Protista. Some algae are **unicellular**, whereas others are **colonial** or **multicellular**. Colonial species live as aggregates of cells that exist attached to one another; however, each cell continues to operate in a largely independent manner. Colonies can be found in many forms and shapes. Some are rounded, forming spherical patterns. Others form linear **filaments**. Colonies comprised of flagellated cells are motile. Others may be sessile and live attached to an underwater substrate. Yet other colonies may be found as nonmotile masses that float in the water column. Filamentous species may grow in **sessile**, submerged forms or may be found in floating masses on the surface of ponds. Multicellular forms are commonly called seaweeds or kelps and are mostly marine.

The term "algae" has also been used to describe **cyanobacteria**, a group of prokaryotic photosynthetic organisms sometimes referred to as blue-green algae. Cyanobacteria grow as filamentous colonies and can be found in many of the same environments that support eukaryotic algae. These are prokaryotic organisms capable of photosynthesis; however, they lack chloroplasts.

Eukaryotic algae have many characteristics in common with plants, including photosynthesis, cell wall components, and use of starch for long-term energy storage. Of all the various types of algae, the green algae are most similar to plants. Green algae may be unicellular, colonial, or multicellular. Regardless, green algae contain chloroplasts with the same photosynthetic pigments found in plant chloroplasts, chlorophyll *a* and chlorophyll *b*. They also produce oxygen as a byproduct of photosynthesis. Multicellular forms superficially resemble plants in their overall structure. They have a stem-like **stipe** that supports flattened **blades** that resemble plant leaves.

Other groups of algae use different combinations of pigments for photosynthesis. These alternative pigments also impact the dominant colors of the different algal groups. We look at each of these in more detail as we survey the world of algae.

Activities

■ Activity 15.1: Division Chlorophyta—The Green Algae

Most green algae are found in freshwater habitats. The group is quite diverse and includes unicellular, multicellular, and colonial forms. Members of Division Chlorophyta are characterized by three common features. Chlorophyll *a* and chlorophyll *b* are the main photosynthetic pigments, although a few species possess other accessory pigments. **Starch** is the primary energy reserve molecule. The cell walls are constructed mainly of **cellulose**. These are the same features that link them closely with plant evolution.

Another similarity some green algae share with plants is an alternation of generations in their life cycle. Some algae have dibionic life cycles in which the alternating generations appear distinct from one another. Others have a **monobionic** life cycle in which the alternating generations look alike and share the same body plan. Many green algae have **isogamous** reproduction, in which both gametes are the same size and shape. Others exhibit **oogamous** reproduction where the sperm is smaller and motile and the egg is larger and stationary.

The simplest body plan for a green alga is unicellular. *Chlamydomonas* is one such species. Each cell is motile by means of two flagella of equal size extending off the apical end of the cell. *Chlamydomonas* cells have a large cup-shaped chloroplast. Inside the chloroplast is a small structure called the **pyrenoid body**. It is involved in starch synthesis. *Chlamydomonas* also has a stigma near the apical end of the cell. The **stigma** is a red-colored structure in the cytoplasm that is used to detect light. The stigma is sometimes called an eyespot. No vision is associated with the stigma; however, the individual cells can detect the presence of light and modify their behavior based on that information.

The life cycle of *Chlamydomonas* is simple. The only diploid cell is the zygote, which immediately divides by meiosis to produce four haploid offspring. These haploid cells continue to divide by mitosis until two come together for sexual reproduction. At that time they shed their flagella, undergo plasmogamy, the fusion of cytoplasms, and karyogamy, the fusion of nuclei, to produce the zygote.

1. Obtain a prepared slide of *Chlamydomonas* or make a wet mount of a living culture.
2. Observe the specimen under low power. You should be able to find the ovoid cells at this magnification. Focus on one individual and switch to high power. Diagram what

you see in the space below. You should be able to label the chloroplast and nucleus. You may be able to see the stigma and pyrenoid body. Although two flagella are present, they may not be visible via light microscopy.

Another common form of green algae is a motile colony. The cells comprising a motile colony are similar to those of *Chlamydomonas* and likely share a common ancestor with *Chlamydomonas*. Division occurs in much the same way as in *Chlamydomonas*: the cells undergo mitotic division; however, instead of being released as in *Chlamydomonas*, the daughter cells of colonial species are retained by a **gelatinous matrix**.

Pandorina is a species of green algae that form motile colonies. Each cell is ovoid in shape and possesses two terminal flagella, much like *Chlamydomonas*.

3. Observe a prepared slide or wet mount of *Pandorina*. Focus on a single colony under high power.

4. Diagram the colony below. You should see a cluster of circular cells. Although each cell possesses two flagella, you may not be able to observe them. Label the daughter cells and gelatinous matrix.

Eudorina is an algal species that forms motile colonies much like *Pandorina*. *Eudorina* is somewhat more complex and forms daughter colonies within the gelatinous matrix of the parent colony. Flagella are present on each cell, but are difficult to see.

5. Label the daughter colonies and gelatinous matrix in **Figure 15-1**.

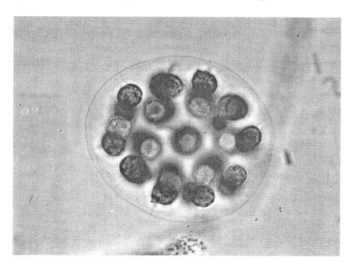

FIGURE 15-1   *Eudorina elegans* in a small motile colony of green algae. (Courtesy of Great Lakes, EPA)

Volvox is the most complex of the colonial forms. Daughter colonies form within the original parent colony through the mitotic division of specialized cells within the colony. The colony is held together by cytoplasmic connections instead of a simple gelatinous matrix. *Volvox* is also capable of sexual reproduction through the production of motile sperm that swim to a female colony to fertilize a nonmotile egg. The zygote is retained within the female colony and develops a thick, resistant cell wall. It is now a **zygospore**. The zygospore remains dormant for a period of time and then divides by meiosis to produce haploid cells that will form a new colony.

6. Obtain a prepared slide of *Volvox*. Note the intricate structure of the colony. Diagram a colony below. Label the parent colony, the daughter colony, and the cytoplasmic connections.

Filamentous forms of green algae are relatively common. Some live attached to underwater substrates through a specialized structure called a holdfast. Others are found floating in the water column or on the surface of standing water. *Spirogyra* is one example of a filamentous green alga.

Spirogyra is commonly found growing in ponds and pools where there is little water movement. When conditions are optimal, photosynthesis can occur so rapidly the oxygen produced gets trapped in the filaments, causing them to rise to the surface of the water.

7. Observe a prepared slide of vegetative *Spirogyra* or a wet mount of living *Spirogyra* under low power. The chloroplasts of *Spirogyra* are arranged in a spiral pattern along the inside surface of the cell. Intermixed with the chloroplasts are pyrenoid bodies. They appear as numerous small dark structures on the chloroplasts.

8. Diagram a filament of *Spirogyra* below. Label the chloroplasts, pyrenoid bodies, cytoplasm, and cell wall.

Spirogyra reproduces sexually and asexually. Asexual reproduction occurs by mitotic division and by fragmentation of the filaments. Sexual reproduction occurs by a specialized process called **conjugation**. In conjugation, two compatible filaments of *Spirogyra* align and form tubular connections between cells in adjacent strands. *Spirogyra* does not produce sperm or egg cells. Instead, the cytoplasm of one cell condenses and moves through the conjugation tube to merge with the cytoplasm of the adjacent cell. The cytoplasm that moves through the conjugation tube is referred to as the **motile gamete**. The recipient cytoplasm is the **nonmotile gamete**. The resultant diploid zygote is retained within the cytoplasm of the second cell. It typically develops a thick, protective outer wall after which it is called a **zygospore**.

9. Obtain a prepared slide of *Spirogyra* conjugating. Observe the specimen under low power. Locate a conjugation tube.

10. Label a motile gamete, nonmotile gamete, and conjugation tube on **Figure 15-2**.

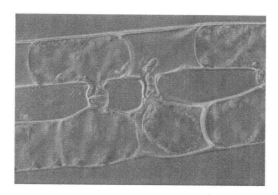

FIGURE 15-2   Compatible filaments of *Spirogyra* have been brought together in culture and are undergoing conjugation. (© M. I. Walker/Photo Researchers, Inc.)

11. Locate a zygospore on your slide. Diagram it along with the cell containing it below.

Oedogonium is a filamentous green alga that exhibits oogamous reproduction. This species forms specialized structures called **gametangia** to produce gametes. Each **oogonium** produces an egg, whereas each **antheridium** produces sperm. The oogonia are larger in diameter than most cells in the filament. Conversely, the antheridia are much smaller than vegetative cells not involved in sexual reproduction. Each oogonium has a fertilization pore that forms in the cell wall to allow the sperm to enter. The egg is fertilized within the oogonium, and the zygote remains within it until it develops into the zygospore.

12. Obtain a prepared slide of *Oedogonium*. Observe the filament under high power. Locate an oogonium. Sketch it below. Does it contain a zygospore?

13. Locate an antheridium, identified by the small size of the cells. Diagram it below.

Multicellular green algae, such as *Ulva* and *Derbesia*, exhibit an alternation of generations. In some cases, such as with *Ulva*, the generations are monobionic. In other cases, as in *Derbesia*, the generations are dibionic. Gametes are isogamous in some and heterogamous in others. Generally, the gametes are motile and free-swimming and the life cycles are quite similar. The gametes fuse in syngamy (fertilization) to produce the diploid zygote. The zygote divides by mitosis to develop into a multicellular sporophyte, which produces spores in sporangia. The haploid spores are free-swimming flagellated cells that lose their flagella when they begin to germinate. Spore germination leads to the production of a multicellular gametophyte that produces gametes in gametangia.

Plants are most closely related to one group of green algae: the charophytes. Because of this common history, plants have a higher degree of similarity with the charophytes than with other algal groups. Charophytes are multicellular seaweeds. They look more like a plant than any of the algae observed so far. These algae exhibit an alternation of dibionic generations, with different body forms in each generation. *Chara* is the genus of charophyte. It is characterized by a body that branches in stem-like patterns. Nodes and internodes can be identified. Branches are attached in whorls. Reproduction in *Chara* is a bit more complex than in other multicellular green algae. Embryophytes, including all plants, produce multicellular reproductive structures that are surrounded by sterile cells. *Chara* shares this characteristic with plants. By this definition of embryophyte, *Chara* could be classified with the plants rather than with the algae.

14. Observe the specimens of multicellular green algae on display. Your instructor will provide you with additional information and directions.

COMPARE AND CONTRAST

1. What characteristics are shared by all the chlorophytes you observed?

2. How are *Volvox* and *Pandorina* similar? How are they different?

3. Compare and contrast *Oedogonium* and *Spirogyra* in terms of structure and methods of reproduction.

■ Activity 15.2: Division Rhodophyta—The Red Algae

The red algae are distinct from other algal groups, and from plants, due to several characteristics. They do not rely on the same pigments for photosynthesis. Rather, they contain the **phycobilin** accessory pigments found in cyanobacteria. Many are reddish in coloration due to the presence of **phycoerythrin**; however, many species appear in other colors due to the addition of other pigments. They do not produce the type of starch found in plants. Instead, excess glucose from photosynthesis is stored as **floridean starch**, a branched molecule reminiscent of glycogen. Floridean starch is stored in cytoplasmic granules rather than in the chloroplast.

Their cell walls contain cellulose like plants and other algae. In addition to cellulose, the cell walls also contain **sulfated galactans** that form a thick mucilaginous layer. These sulfated galactans are used commercially as thickening and stabilizing agents in a variety of foods, including salad dressings and ice cream. Another important product harvested from red algae is **agar**. Agar is a solidifying agent used in the preparation of solid microbiological culture media.

1. Observe the red algal specimens on display. Many are highly branched and filamentous in appearance. Coralline red algae have walls with large deposits of calcium carbonate, making their bodies hard but brittle.

2. Your instructor will provide additional directions concerning the specimens on display.

COMPARE AND CONTRAST

1. How are the red algae on display similar to the multicellular green algae?

....

....

....

....

....

2. How can you describe the differences between coralline red algae and other members of this group?

....

....

....

....

....

■ Activity 15.3: Division Phaeophyta—The Brown Algae

The brown algae are multicellular forms that live mostly in marine environments and are quite common in intertidal zones. They include the kelps. Brown algae are photosynthetic through the action of chlorophyll *a*, chlorophyll *c*, and a variety of accessory pigments. They do not store excess glucose as starch but rather as mannitol, fats, or **laminarian**, a polymer of glucose.

1. Observe a preserved specimen of *Fucus*, a common intertidal brown alga. Sketch a representative section of the body of *Fucus* below. How would you describe its form?

Kelps are large forms of brown algae with bodies composed of three parts: the **hold-fast**, which anchors the body to a substrate; a **stipe**, which forms a vertical stalk; and flattened **blades** for photosynthesis. Some kelps form air bladders, or **pneumatocysts**, on the blades to help keep them afloat in the water column. In some species these are small, bubble-like structures on the surface of blades, whereas in other they are large, central structures to which blades attach.

2. Observe the specimens on display. Sketch a typical kelp below and label the holdfast, stipe, and blades.

3. Are pneumatocysts present? If so, describe their appearance.

COMPARE AND CONTRAST

1. Compare and contrast *Fucus* and kelps.

2. How are kelps similar in structure to the multicellular green algae? How are they different?

■ Activity 15.4: Division Chrysophyta—Diatoms and Their Relatives

Division Chrysophyta includes the unicellular diatoms and the multicellular golden-brown and yellow-green algae. In this lab we focus on the diatoms. Additional information about the yellow-green algae and the golden-brown algae can be found elsewhere.

Diatoms are unicellular algae classified in the same division as the brown algae. Diatoms have a distinctive cell wall made up of two halves (**frustules**) that fit together in a box-and-lid fashion. The walls contain silica, the main ingredient in glass, and often exhibit intricate patterns on the surfaces of the walls. When the diatoms die, their frustules sink to the bottom of the ocean and form diatomaceous earth. **Diatomaceous earth** is used in a number of commercial products, including filters, abrasives, and reflec-torized paint.

Reproduction in diatoms is somewhat more complicated than in other unicellular algae. When a cell divides mitotically, each daughter cell receives one frustule and forms a second frustule to fit slightly inside of the original. Over generations, some of the diatoms get progressively smaller. They continue this pattern of asexual reproduction and decreasing size until a critical minimum size is reached. This stimulates the diatom to divide meiotically, producing gametes that fuse in syngamy to produce a diploid zygote. The zygote then divides mitotically and the process repeats.

1. Obtain a prepared slide of mixed diatoms and view them under high power. Identify four or five examples of diatoms. Diagram them in the space below. This is a mixed slide. Other forms of algae will likely be present. Diatoms look like small, intricate glass boxes when viewed microscopically.

COMPARE AND CONTRAST

1. Why are the frustules of diatoms important in a commercial sense?

....

....

....

....

....

2. How do diatoms compare with unicellular green algae, such as *Chlamydomonas*?

....

....

....

....

....

■ Activity 15.5: Division Pyrrhophyta—The Dinoflagellates

The dinoflagellates are a unique group of unicellular algae for many reasons. They possess flagella; however, one flagellum extends outward from the cell, whereas the other lies in a groove that wraps around the middle of the cell. Most species are photosynthetic due to chlorophyll *a* and chlorophyll *c* as well as some unusual cartenoid pigments. Some species of dinoflagellates are completely heterotrophic.

Dinoflagellates are a concern to human health because of the phenomenon known as a "**red tide**." These are essentially algal blooms that occur when oceanic conditions are conducive to rapid reproduction of these dinoflagellates. Some species are red-pigmented, hence the name "red tide"; however, other species contain different pigments. Dinoflagellate blooms can be dangerous even if the water never appears red due to the high numbers of algae present. The danger to humans is due to toxic substances that are produced as a byproduct of the algae's metabolic activities. These toxins accumulate in shellfish in high concentrations. Shellfish are filter feeders and acquire their food directly from any organic material available in the water. When humans ingest contaminated shellfish, they can acquire a disease called paralytic shellfish poisoning caused by the neurotoxins that have accumulated in the tissues of the shellfish.

1. Obtain a prepared slide of mixed dinoflagellates. Observe the specimens under high power. Diagram two or three examples in the space below.

COMPARE AND CONTRAST

1. Explain the structural differences between diatoms and dinoflagellates.

....

....

....

....

....

2. Both dinoflagellates and *Chlamydomonas* have two flagella. How do their flagellar arrangements compare with one another?

....

....

....

....

....

▪ Activity 15.6: Division Euglenophyta—The Euglenoids

The euglenoids are a diverse group of unicellular, flagellated protists. Some do not possess chloroplasts and are considered to be protozoans by zoologists. The genus *Trypanosoma*, for example, is a parasitic euglenoid that lives in the bloodstream. Members of this genus cause African sleeping sickness and Chagas' disease. The cells of *Trypanosoma* are quite similar to those of other euglenoids, with the exception of no chloroplasts.

Euglena is an example of a genus that does contain chloroplasts and is photosynthetic. The cells are elliptical in shape. Multiple chloroplasts are present in the cytoplasm. The plasma membrane is associated with a semirigid, elastic layer of proteins to form a structure called the **pellicle**. They do not possess a cell wall.

1. Obtain a prepared slide of *Euglena* or make a wet mount of living *Euglena* sp. for observation.
2. Observe the specimen under high power. Diagram a representative cell below. The chloroplasts of *Euglena* contain a food reserve molecule called **paramylum** rather than starch. The paramylum granules appear as shiny granules in the cell. The stigma is a small, reddish, light-sensitive structure near the end of the cell containing the single flagellum. Label the pellicle, stigma, chloroplasts, cytoplasm, and paramylon granules (if visible).

COMPARE AND CONTRAST

1. How is *Euglena* similar to *Chlamydomonas*? What notable differences separate these two genera?

....

....

....

....

....

2. Compare *Euglena* to dinoflagellates. What are the most outstanding differences in their cell structure?

....

....

....

....

....

▣ Activity 15.7: Division Cyanobacteria—The Blue-Green Algae

Cyanobacteria are not true algae, although they are often confused with filamentous algae. These are prokaryotic organisms in Domain Bacteria. They grow as filamentous colonies and can often be found growing in freshwater pools alongside filamentous green algae, such as *Spirogyra*. As bacteria, these are unicellular organisms that contain a cell wall composed of peptidoglycan, a complex carbohydrate polymer. They are photosynthetic; however, they do not contain chloroplasts or other membrane-bound organelles. Photosynthesis occurs in the cytoplasm and at the plasma membrane in these organisms. Reproduction is through asexual means only.

Nostoc is a common filamentous cyanobacterium that exhibits some specialization of cells. This bacterium is capable of **nitrogen fixation**, capturing gaseous nitrogen from the atmosphere and converting it into forms such as ammonia. It is also photosynthetic and produces oxygen as a byproduct of photosynthesis, as do plants. Cyanobacteria are the only organisms capable of both processes. Nitrogen fixation can only occur in the absence of oxygen, so the filaments include cells specialized for nitrogen fixation. These cells are called **heterocysts**. They are slightly larger in diameter than other cells in the filament and are relatively thick-walled. Microscopically, they generally appear darker in color and often have a brown appearance. Photosynthetic cells in the filament are bluish-green in color.

1. Obtain a prepared slide of *Nostoc*. Observe the specimen under high power. Locate the heterocysts within the strand.

2. Label a heterocyst and vegetative cell on **Figure 15-3**.

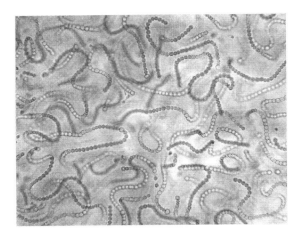

FIGURE 15-3   *Nostoc* filaments with heterocysts (×300). (Courtesy of James D. Mauseth)

3. Observe a preserved specimen of *Nostoc*. These filaments typically form ball-shaped "colonies" in nature. These are not colonies in the true sense. They are basically masses of filaments encased in a large amount of gelatinous material.

4. Diagram a *Nostoc* colony below.

Oscillatoria is another species of filamentous cyanobacteria. It can be found growing in still fresh water or on the surface of damp soil. The cells of *Oscillatoria* are short and relatively box-shaped.

5. Observe a prepared slide of *Oscillatoria* under high power. Diagram a few cells below.

COMPARE AND CONTRAST

1. Compare and contrast the strands of *Oscillatoria* and *Nostoc*.

    ....

    ....

    ....

    ....

    ....

2. How do the filamentous cyanobacteria differ in appearance from the filamentous green algae?

    ....

    ....

    ....

    ....

    ....

Study Guide

*   Be able to define the terms in bold.
*   Be able to identify the algal species studied.
*   Be able to explain the differences between the major groups of algae.
*   Be able to list the key characteristics of all true algae.

Conclusions

1. Describe the colonial, filamentous, and multicellular body plans of algae.

    ....

    ....

    ....

2. Why are the blue-green algae not considered true algae?

    ....

    ....

    ....

3. What characteristics are used for classifying algae?

4. How does *Euglena* differ from *Chlamydomonas*?

5. Describe the process of conjugation in *Spirogyra*.

6. Compare and contrast monobionic and dibionic life cycles.

7. Why are dinoflagellates important to humans?

8. In what way are diatoms useful to humans?

9. Which algae produce paramylum? What is the plant equivalent to paramylum?

10. List the general characteristics shared by all true algae.

Chapter 16

© Cristi Matei/ShutterStock, Inc.

Nonvascular Plants

Laboratory Activities

Activity 16.1: Moss Gametophores

Activity 16.2: Moss Sporophytes

Activity 16.3: The Complete Moss Plant

Activity 16.4: Liverwort Gametophytes

Activity 16.5: Liverwort Sporophytes

Activity 16.6: Hornworts

Goals

Following this exercise students should be able to

- Describe the reproductive pattern of nonvascular plants.
- Identify gametangia of moss and liverwort gametophytes.
- Identify the parts of moss and liverwort sporophytes.
- Understand the importance of the gametophyte and the sporophyte in nonvascular plant reproduction.
- Describe the differences between mosses, liverworts, and hornworts with respect to gametophyte and sporophyte appearance.

Introduction

The nonvascular plants are the most primitive of land plants and have no vascular tissues for transporting water and nutrients (i.e., no xylem or phloem). Several distinctive features set the nonvascular plants apart from the rest of the plant kingdom: absence of xylem and phloem, dominant gametophytes, and dependence on environmental water for reproduction (a characteristic shared with the seedless vascular plants).

The nonvascular plants include three divisions: **Bryophyta** (mosses), **Hepatophyta** (liverworts), and **Anthocerotophyta** (hornworts). The members of each division are quite distinct from one another but share many characteristics, such as the lack of vascular tissues, dominant gametophytes, and the absence of true roots. At one time because of these common characteristics all nonvascular plants were classified in division Bryophyta. Because of this the term **bryophyte** is still sometimes used to refer to a nonvascular plant.

The reproductive cycle of bryophytes and all other plants includes an **alternation of generations**. The two generations are the diploid, spore-producing sporophyte and the haploid, gamete-producing gametophyte. Due to this alternation of generations two types of haploid reproductive cells are needed to complete the plant life cycle: the **spore** and the **gamete**. The spore is the immediate product of meiosis, which occurs in the sporophyte. The gamete is produced later by mitosis within the gametophyte. In nonvascular plants the spore is the primary mechanism of dispersal. When spores are released, they are carried away from the parent plant, usually by wind. If the spore lands in a suitable habitat it germinates, producing a gametophyte. The **gametophyte** produces gametes (eggs and sperm cells). The eggs remain imbedded within gametophyte tissues. They are fertilized in place and then grow into the **sporophyte**. Thus, sporophytes remain attached to the gametophyte, and the spores it produces allow the species to disperse to new habitats.

The life cycle of a typical moss is shown in **Figure 16-1**. We take a closer look at the structure of the gametophyte and sporophyte in mosses and in other nonvascular plants below. Although the structure of the plants varies somewhat, the reproductive cycle is essentially the same.

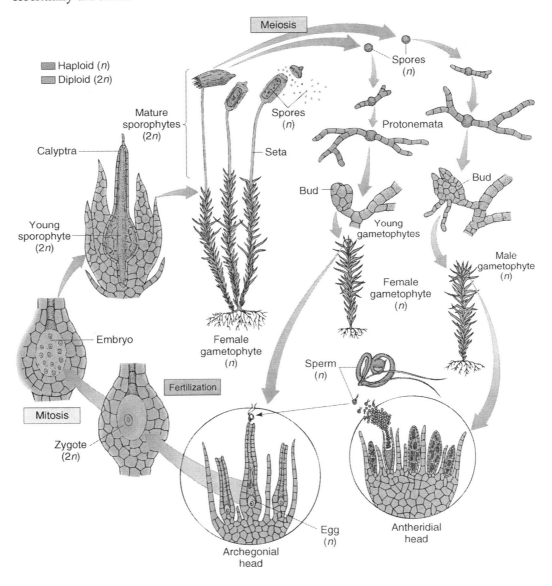

FIGURE 16-1 Life cycle of a moss. (© Jones & Bartlett, LLC)

Activities

■ Activity 16.1: Moss Gametophores

Mosses are the best known of the nonvascular plants. They occupy many habitats worldwide. Mosses are low-growing plants. Their relatively small size is in part due to the lack of vascular tissue. Most mosses have no type of cells involved in transporting water and nutrients. Without efficient transport virtually the entire plant must be photosynthetic to produce sufficient amounts of glucose and other nutrients to sustain it. Likewise, with no water transport system, most mosses rely on the passive movement of water over their outer surfaces to supply the needed moisture to their cells. Mosses are anchored to their substrates by **rhizoids**. Rhizoids are short, root-like structures made of one or a few cells that anchor the plant. They do not function in water or mineral uptake.

1. Observe an intact moss plant. How large is the plant (estimate the height)?

....

....

....

2. Describe the distinctive features of the plant.

....

....

....

Moss gametophytes produce leafy stems called **gametophores**. These gametophores are the form of the plant that is readily recognizable and comprise the gametophyte. Some of these gametophores have **antheridia** at their apical ends. Antheridia are one type of gametangium. **Gametangia** are multicellular structures that produce gametes. In this case antheridia produce sperm cells. The apical end of the gametophores contains multiple antheridia and is called an **antheridial head**.

In the antheridial head shown in **Figure 16-2**, locate an antheridium. They appear as elongated oval structures surrounded by a light "halo" composed of **sterile jacket cells**. Inside the antheridium you will find the **spermatogenous tissue**, the tissue that produces the sperm cells. Interspersed between the antheridia are long strands of sterile cells called **paraphyses**. The paraphyses help hold water on the apical surface of the antheridial head. As water builds up in this area, sperm are released into the water and are therefore able to be splashed to adjacent gametophores as additional raindrops hit the antheridial head.

3. On Figure 16-2 label the antheridium, spermatogenous tissue, sterile jacket cells, and paraphyses.

Moss sperm cells are flagellated, which allows them to actively seek an egg to fertilize by swimming through water. Water is essential for reproduction in nonvascular plants. The sperm are splashed out of the antheridial head by drops of rainwater. Some of those sperm may be splashed onto the top of a female gametophore, whereas others may simply be splashed onto the ground. The apical end of a female gametophore houses the **archegonia** and is thus referred to as the **archegonial head**. Archegonia are the second type of gametangia and function to produce and house the egg. In the archegonial head in **Figure 16-3** find an archegonium. The archegonia appear as vase-shaped structures on the apical side of the archegonial head. Each archegonium contains two regions. The lower **venter** is the swollen end of the archegonium that houses the egg. Extending upward from the venter is an elongated **neck**. The sperm reach the egg by entering through

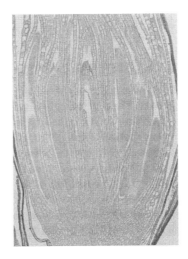

FIGURE 16-2   Microgametangia of moss. (Courtesy of James D. Mauseth)

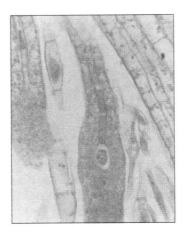

FIGURE 16-3   Megagametangium of moss. (Courtesy of James D. Mauseth)

the **neck canal**, an open channel formed in the center of the neck. The archegonial head also has numerous paraphyses.

4.   On Figure 16-3 label the archegonium, venter, neck, and paraphyses.

COMPARE AND CONTRAST

1.   How do the slides you examined compare with the images above?

....

....

....

....

....

2. How does the structure of the antheridial head differ from that of the archegonial head? How do these differences in structure relate to the reproductive cycle of the moss?

....

....

....

....

....

■ Activity 16.2: Moss Sporophytes

Some of the gametophores on display (living or persevered) have sporophytes attached to them. The sporophyte grows out of the apical end of a female gametophore where the egg was fertilized. Each sporophyte is distinctive. The spore-producing region of the sporophyte is called a **sporangium**. In mosses this sporangium is referred to as a **capsule**. The capsule is supported by an elongated stalk called the seta. At the base of the seta is a swollen end called the **foot**. The foot functions to attach the sporophyte to the gametophyte. It is also responsible for obtaining water and nutrients from the gametophyte tissues. The capsule is coverd by a lid called the **operculum**. The operculum is roughly cone-shaped and breaks off of the capsule when the spores are mature. Just inside the opening left by the operculum is a row of teeth-like structures called the **peristome**. The "teeth" of the peristome bend in response to changes in humidity and aid in spore dispersal. During early stages of sporophyte development, the capsule is covered by a portion of gametophyte tissue derived from the archegonium. This is the **calyptra**. The calyptra falls off of the operculum end of the capsule before spore release.

1. Diagram a moss sporophyte in the space below. Label the capsule, foot, seta, and operculum.

2. Obtain a slide of a longitudinal section of the moss capsule.
3. You should be able to find the operculum and peristome at one end of the capsule. The operculum is attached to the walls of the capsule by slightly enlarged **dehiscent cells**. As these dehiscent cells lose water through evaporation, they break open and allow the operculum to separate from the capsule. A central sterile band of cells, the **columnella**, extends through the central portion of the capsule. The columnella is surrounded on either side by the forming spores.

4. In the space below, draw a longitudinal section of a moss capsule. Label the operculum, peristome, columnella, spores, capsule, and dehiscent cells.

COMPARE AND CONTRAST

1. Is the sporophyte independent of or dependent on the gametophyte for nourishment? How can you determine this based on the structure, appearance, and color of the gametophyte versus the sporophyte?

....

....

....

....

....

2. How does the structure of the archegonial head relate to the location of the sporophyte?

....

....

....

....

....

▪ Activity 16.3: The Complete Moss Plant

Using the information above, label the following structures on **Figure 16-4**: rhizoids, gametophore, seta, capsule, foot, operculum, sporophyte, and gametophyte.

FIGURE 16-4   The complete moss plant. (© Jones & Bartlett, LLC)

▪ Activity 16.4: Liverwort Gametophytes

Liverworts are small plants that live in moist habitats. There are two types of liverworts: **leafy** and **thallose**. Leafy liverworts look much like a moss. The main body of a thallose liverwort gametophyte is a flattened ribbon-like structure referred to as the **thallus**. We focus on the thallose liverworts using *Marchantia* as our example organism.

Liverworts can reproduce sexually and asexually. Asexual reproduction is mediated by the production of **gemmae** (singular = gemma). Gemmae are multicellular structures produced on the surface of the thallus in cup-shaped structures called **gemmae cups**. When raindrops hit a gemmae cup, some of the gemmae splash out and can then germinate and grow into another liverwort gametophyte.

1. Observe a live or preserved specimen of *Marchantia* with gemmae cups.

2. Notice the distinctive shape of the gemmae cups. How many gemmae cups are present on the specimen provided?

3. Observe a prepared slide of a longitudinal section through a *Marchantia* gemmae cup.

4. In the space below, sketch a region of the thallus containing a gemma cup. Label the gemma, thallus, and gemma cup.

Marchantia is **dioecious**. Each gametophyte produces either archegonia or antheridia. Both types of gametangia are produced as vertical outgrowths referred to as archegoniophores and antheridiophores, respectively. The **archegoniophore** is a vertical stalk with numerous finger-like projections at the apical end that drape downward. The underside of the archegoniophore houses the **archegonia**, which are responsible for producing the egg. The **antheridiophore** is an umbrella-shaped vertical outgrowth responsible for producing sperm in the **antheridia** that cover its upper surface. The antheridia and archegonia of *Marchantia* are similar in structure to those of the mosses.

5. Label the antheridiophore and the antheridiam on **Figure 16-5**.

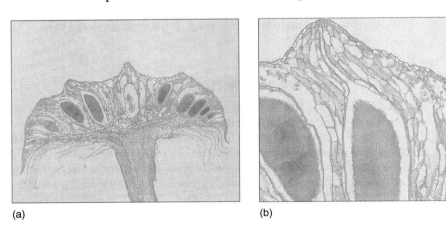

(a)   (b)

FIGURE 16-5   *Marchantia* antheridiophore with antheridia (a, ×8; b, ×32). (Courtesy of James D. Mauseth)

6. Observe a prepared slide of a *Marchantia* antheridiophore. Diagram the apical end of the antheridiophore in the space below. Label the antheridia, spermatogenous tissue, and sterile jacket cells.

7. Observe a prepared slide of a *Marchantia* archegoniophore. Label an archegonium, neck, venter, and egg on **Figure 16-6**.

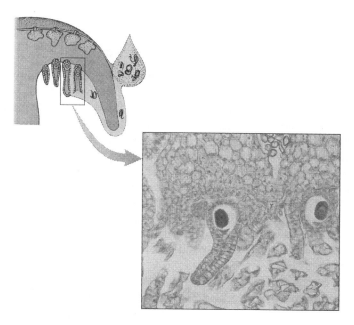

FIGURE 16-6   *Marchantia* archegoniophore with archegonia. (Courtesy of James D. Mauseth)

COMPARE AND CONTRAST

1. How are the gametophytes of liverworts similar to those of mosses? How are they different?

....

....

....

....

....

2. Are the liverwort gametophytes haploid or diploid? How does this relate to the mechanism by which gametes are produced?

....

....

....

....

....

■ Activity 16.5: Liverwort Sporophytes

The eggs remain inside the venter while fertilized. The sperm are flagellated and swim through water surrounding the liverworts until they reach an archegonium. The sperm then enter the archegonium through the neck and fertilize the egg within the venter. Once fertilized the egg becomes a zygote, which begins growing into the sporophyte plant.

The *Marchantia* sporophyte lives permanently attached to the gametophyte. It attaches to the gametophyte through the archegonium in which the egg was originally produced. The sporophytes form a cluster on the lower surface of the archegoniophore. *Marchantia* sporophytes have the same basic structure as the moss sporophyte seen above; however, it is a more simplified form. The sporophyte is attached to the gametophyte by its **foot**. The foot also acquires nutrients and water from the gametophyte plant. A short, stocky **seta** supports the large, elliptical **sporangium**. Spores are formed by **sporocyte** cells undergoing meiosis within the sporangium. The gametophyte frequently produces a thin flap of tissue, originally part of the archegoniophore, that temporarily covers the growing sporophyte. This is the **calyptra**. Within the sporangium surrounded by the newly formed spores, you can see some diploid, elongated cells called **elators**. These elators have spirally thickened cell walls that uncoil when the sporangium opens and help to disperse the spores.

1. Observe a prepared slide of a *Marchantia* sporophyte.
2. Label the sporangium, foot, seta, and archegoniophore on the sporophyte on **Figure 16-7**.

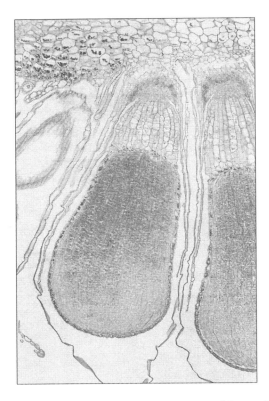

FIGURE 16-7   Liverwort sporophyte. (Courtesy of James D. Mauseth)

COMPARE AND CONTRAST

1. How does the sporophyte of the moss differ from that of the liverwort? How are they similar?

2. In both mosses and liverworts, why are sporangia raised above the surface of the gametophyte?

3. Compare the structures responsible for this in the two types of plants.

4. How does the gametophyte of a thallose liverwort differ from that of a moss in appearance?

▦ Activity 16.6: Hornworts

Hornworts are small, thallose plants that live in moist areas. The gametophyte thallus is similar to that of the thallose liverworts. Reproductive structures and mechanisms are also similar to thallose liverworts. The differences between the gametophytes of thallose liverworts and hornworts are seen at the cellular level. Hornworts do not contain the oil bodies common in liverworts. Hornwort cells are distinct from all other plant cells in containing a single chloroplast.

Hornworts are named after the appearance of their sporophytes, which grow as tall, cylindrical structure rising from the gametophyte tissues. The sporophyte is attached to the gametophyte by the sporophyte's **foot**; however, no seta is present. Instead, the remainder of the sporophyte grows upright from the foot. The diameter is relatively constant, and growth is continuous from a **basal meristem**.

The hornwort sporophyte is green and capable of photosynthesis. As such, it is the only sporophyte of nonvascular plants that is only partially dependent on the gametophyte. It is still incapable of independent growth, but some of its nutrients are provided by photosynthesis within the sporophyte. The sporophyte contains a **columnella**, similar to that seen in mosses. Surrounding the central columnella is sporocytes and newly formed spores (farther from the basal meristem). **Elators** are found within the sporophyte. These elongated, multicellular structures are imbedded within the chamber housing the spores.

1. On **Figure 16-8** label the sporophyte, gametophyte thallus, foot, and basal meristem.

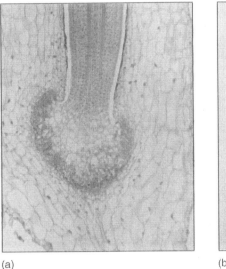 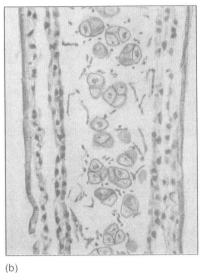

(a)   (b)

FIGURE 16-8   Hornwort sporophyte. (Courtesy of James D. Mauseth)

COMPARE AND CONTRAST

1. Are hornwort sporophytes more similar to liverwort or moss sporophytes? Why?

....

....

....

....

....

2. How do the elators of hornworts differ from those of liverworts?

Study Guide

- Be able to define the terms in bold.
- Be able to identify the structures labeled in the drawings.
- Be able to describe the reproductive cycles of the three nonvascular plant groups and structures involved.
- Be able to answer each of the questions asked in the lab exercise.

Conclusions

1. Why is it important for the sporangia to be raised above the surface of the parent gametophyte plant?

2. Why are these plants considered to be nonvascular?

3. Explain why the term "bryophyte" is technically incorrect when referring to all non-vascular plants collectively.

4. Describe the role of elators in liverworts and hornworts.

5. List and describe the two major types of gametangia and what type of gamete is produced by each.

6. Correctly identify the following structures as haploid or diploid.
   a. Archegonium

   b. Sporangium

   c. Sperm

   d. Spore

   e. Elator

   f. Calyptra

   g. Seta

   h. Antheridiophore

7. Describe the primary means of dispersal for nonvascular plants. What cells are involved in dispersal?

8. Which generation is dominant in the nonvascular plants, the sporophyte or the gametophyte?

9. Describe the alternation of generations as it occurs in the nonvascular plants.

10. What is the primary function of gemmae? How is this beneficial to a plant?

Chapter 17

Seedless Vascular Plants

Laboratory Activities

Activity 17.1: Division Lycophyta

Activity 17.2: Division Arthrophyta

Activity 17.3: Division Pteridophyta

Activity 17.4: Wisk Ferns

Goals

Following this exercise students should be able to

- Describe the life cycle of the seedless vascular plants.
- Define the reproductive structures of the seedless vascular plants.
- Characterize the different groups of seedless vascular plants based on anatomy.
- Explain the role of spores in the life of seedless vascular plants.

Introduction

The seedless vascular plants are a diverse group of plants that include ferns, horsetails, club mosses, and quillworts. These plants are commonly referred to as ferns and "fern allies" or as vascular cryptograms. The fern allies have many physical distinctions from the ferns, but the reproductive strategies are the same for all of them. Seedless vascular plants rely on spores as their primary means of dispersal. Reproductive structures are somewhat similar to those observed in the nonvascular plants; however, the seedless vascular plants have a major advantage over the nonvascular plants: the production of true vascular tissues, xylem and phloem. Vascular tissues are characteristic of the sporophytes of the seedless vascular plants, gymnosperms, and angiosperms. In all these types of plants the sporophyte is the dominant generation. The gametophytes exhibit varying degrees of dependency on the sporophyte generation and do not contain xylem or phloem.

There are two main lineages of seedless vascular plants: the **microphyll** and **megaphyll** lineages. Microphylls are leaves that are not associated with leaf gaps. They originated as enations, leaf-like outgrowths of stem tissue that increased the photosynthetic surface of the plant. Division Lycophyta contains the only extant plant species with microphylls.

The sporangia of lycophytes are produced on the surface of modified leaves called **sporophylls**. In many lycophytes, sporophylls are produced in cone-like clusters called strobili (singular = strobilus). Some species are **homosporous** and produce only one type of spore. Others are **heterosporous** and produce both microspores and megaspores. The spores germinate to produce gametophytes. In some lycophytes the gametophytes develop almost entirely within the walls of the spore.

Megaphylls are leaves that are associated with the formation of leaf gaps. In general, these leaves exhibit a more complex structure than microphylls. Megaphylls are characteristic of many of the seedless vascular plants, including the numerous species of ferns. They are also found in gymnosperms and angiosperms. Overall, megaphylls are the more common type of leaf in Kingdom Plantae.

Activities

■ Activity 17.1: Division Lycophyta

Lycophytes possess true roots and leaves. Although the modern species of lycophytes are fairly small plants, some extinct species were large trees with leaves reaching approximately 1 m in length. Today, extant lycophytes can be found in three common genera: *Lycopodium*, *Selaginella*, and *Isoetes*. Each shares the same reproductive strategy, relies on spores for dispersal, and exhibits the alternation of generations characteristic of all plants.

Lycopodium is a genus of homosporous plants that are commonly called ground pines or club mosses. These plants can be found growing on forest floors and superficially resemble the shape of small pine trees. These plants possess true root and leaves. The spores are produced within sporangia clustered into strobili. The strobili are found on the tips of stems. Spores are released into the air for wind dispersal.

Each sporangium in the strobilus of *Lycopodium* contains numerous **sporocyte** cells that divide by meiosis to produce numerous equal-sized spores. The plant is, therefore, homosporous. Each spore is released and carried by air currents some distance from the parent plant. Spores germinate to produce the gametophyte plant. The gametophyte produces the gametes (egg and sperm). The fertilized egg (zygote) grows into a new sporophyte plant.

1. Observe the living or preserved specimen on display. How would you describe the microphylls of this plant?

....

....

....

2. Describe the overall growth pattern of *Lycopodium*.

....

....

....

3. Obtain a prepared slide of the longitudinal section of a *Lycopodium* strobilus.
4. Observe the specimen under scanning power.
5. Notice the central axis running lengthwise through the strobilus. Multiple sporophylls can be seen branching off of this central axis. Each sporophyll supports an elliptical sporangium full of developing spores.

6. Draw a section of the *Lycopodium* sporangium. Label a sporophyll, spore, and sporangium.

Selaginella is a heterosporous lycophyte. The microphylls of *Selaginella* are similar to those of *Lycopodium*; however, the plants are typically low-growing and do not resemble miniature pines. As a heterosporous plant *Selaginella* produces both **megaspores** and **microspores**. Megaspores are larger than microspores and germinate to produce the female gametophyte. In contrast, microspores are smaller and germinate to produce the male gametophyte. Megaspores are produced from a meiotic division of **megasporocyte** cells within a megasporangium. The **megasporangium** is found on the surface of a **megasporophyll** that is typically found in a strobilus. Likewise, microspores are produced from the meiotic division of **microsporocytes** within a microsporangium. **Microsporophylls** support the **microsporangia** and are found in strobili along with megasporophylls.

1. Observe the living or preserved specimen on display. How would you describe the microphylls of *Selaginella*?

....

....

....

2. Describe the overall growth pattern of this plant.

....

....

....

3. Obtain a prepared slide of the longitudinal section of a *Selaginella* strobilus.
4. Observe the arrangement of megasporophylls and microsporophylls. Are they found on the same side of strobilus?

....

....

....

5. Draw a section of the *Selaginella* strobilus below. Label the megasporophyll, megasporangium, megaspore, microsporophyll, microspore, and microsporangium.

Isoetes is a genus of lycophyte that has large microphylls that grow from corm-like stems. The plants are heterosporous, and the sporangia are housed in the bases of the leaves. These plants are commonly called quillworts and can be found growing in areas that are generally wet or muddy and seldom dry out.

1. Observe the living or preserved specimens of *Isoetes* on display. How would you describe the leaves of this plant?

....

....

....

COMPARE AND CONTRAST

1. Describe the characteristics that are common to all the lycophytes.

....

....

....

....

2. Compare and contrast *Lycopodium* and *Selaginella* with reference to reproductive structures, types of spores produced, and overall appearance.

....

....

....

....

....

3. How are *Isoetes*, *Lycopodium*, and *Selaginella* similar?

....

....

....

....

....

▪ Activity 17.2: Division Arthrophyta

Members of Division Arthrophyta are commonly called equistophytes. They are representatives of the megaphyll lineage of seedless vascular plants. The most familiar of the equistophytes are found in genus *Equisetum* and commonly called horsetails or scouring rushes. The stems contain ridges that have a coarse texture, which allowed them to be used by pioneers to clean pots and pans.

Equisetum produces sporangia on modified leaves called **sporangiophores**. Each sporangiophore contains up to 10 sporangia. The sporangiophores are attached to a central cone axis to form a strobilus. The outer surface of the strobilus is formed by adjacent sporangiophores, the top of which are octagonal in shape.

The strobili are found on the apical tips of green, photosynthetic stems. Scale-like leaves, often brown in color, are found attached to the nodes of the stem. Nodes are separated by fairly long internodes.

1. Observe the living or prepared specimens on display.
2. Diagram a portion of the stem including a scale leaf below. Label the **node, internode,** and **leaf.**

3. On **Figure 17-1** label the strobilus and sporangiophores.

Equisetum stems contain a large number of canals (**Figure 17-2**). The large **central canal** is formed by pith breaking down. Between the central canal and the edge of the stem are the **carinal canals**. The carinal canals are found opposite the ridges of the stem within the vascular bundles. **Cortical canals** are found in the cortex of the stem and are found within the area encompassed by the ridges of the stem.

4. Obtain a prepared slide of the cross-section of an *Equisetum* stem.
5. Observe the specimen under scanning power. Notice the canals found throughout the stem.

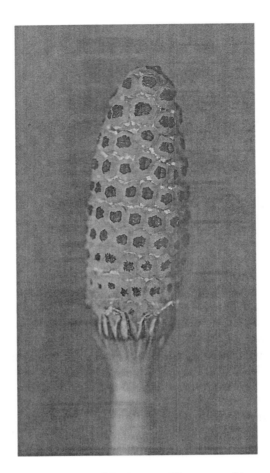

FIGURE 17-1   Sporangium of *Equisetum*. (Courtesy of James D. Mauseth)

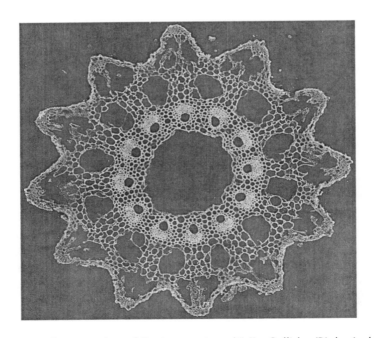

FIGURE 17-2   Cross-section of *Equisetum* stem. (© Jim Solliday/Biological Service)

6. Draw the cross section of the stem and label the **epidermis**, carinal canal, cortical canal, and central canal.

7. Obtain a prepared slide of the longitudinal section of *Equisetum* sporangiophores or strobilus.

8. Observe the specimen under scanning power. Note the central axis to which all of the sporangiophores are attached. The sporangiophores extend outward into a flattened area that forms the outer covering of the strobilus. This flattened outer covering protects the elliptical sporangia.

9. Draw a portion of the *Equisetum* strobilus. Include one intact sporangiophore. Label the sporangiophore, central axis, sporangium, and spores with elators.

Equisetum spores are specialized for wind dispersal by possessing external flaps of tissue called **elators**. These elators are haploid, as is the entire spore. They aid in spore dispersal by increasing the area available to catch the wind.

10. Obtain a prepared slide of *Equisetum* spores.

11. Draw and label a spore with elators below.

COMPARE AND CONTRAST

1. Compare the structure and function of horsetail elators with the elators of liverworts and hornworts.

....

....

....

....

....

2. How does the internal structure of *Equisetum* strobili differ from that of *Lycopodium* or *Selaginella*? How are they similar?

....

....

....

....

....

▪ Activity 17.3: Division Pteridophyta

Ferns are plants with megaphylls that are large and highly subdivided. The megaphylls of ferns are commonly referred to as **fronds**. The fronds are composed of a lamina and a petiole. The petiole attaches to the **rhizome**, a horizontal underground stem. Most fronds are pinnately compound, and the lamina is divided into small leaflets called **pinnae** (singular = pinna). The midrib extending outward from the petiole is the rachis. The pinnae may be further subdivided into **pinnules**, forming a bipinnately compound leaf.

Fern leaves go through dramatic changes as they develop. They emerge from primordia as coiled structures called **fiddleheads** (**Figure 17-3**). As they continue to grow and develop, the fiddleheads unfurl into the familiar frond.

FIGURE 17-3   Fern fiddlehead. (Courtesy of Dr. Shu-Chuan Hsiao, National Chunghsing University)

1. Observe the fern specimens on display.
2. Label the rachis, petiole, lamina, pinna, pinnule, and frond on the leaves shown in **Figure 17-4**.

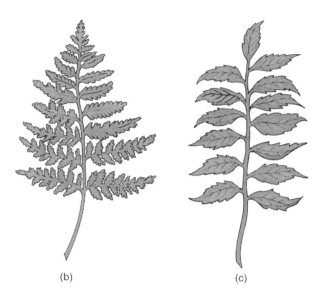

(b) (c)

FIGURE 17-4   Fern fronds. (© Jones & Bartlett, LLC)

The internal arrangement of tissues within the stems of ferns varies from species to species. Each contains xylem and phloem, with xylem typically found in the innermost region of the vascular tissues. Pith may be present in some species. Some rhizomes have an endodermis surrounding the vascular bundles as well.

3. Obtain a prepared slide of a cross-section of *Polypodium* rhizome.

4. Observe the specimen under low power. Locate a vascular bundle. Observe the vascular bundle under high power.

5. Is a pith present in this stem?

....

....

....

6. Diagram a vascular bundle of *Polypodium* below. Label the xylem, phloem, and endodermis.

Ferns produce sporangia in clusters found on the lower sides of their fronds. These clusters are called **sori** (singular = sorus). Sori are found in large numbers on the lower sides of fern fronds. In most ferns leaves function in photosynthesis as well as production of sporangia. The sori are sometimes covered by an outgrowth of tissue called the **indusium**. The indusium helps to protect the sporangia within each sorus. One form of indusium involves the production of an umbrella-shaped structure that grows out from the lower side of the fern frond. Other types of indusia may encircle the sorus but leave the crest of the sorus open to the environment.

7. Obtain a prepared slide of the cross-section of a fern leave with indusium.

8. Observe the specimen under scanning power. Locate a sorus.

9. Diagram a sorus and the associated leaf section below. Label the sorus, indusium, and sporangium.

The sporangium of a fern is a highly specialized structure. Sporangia are rounded, roughly spherical structures that produce spores within the sphere. On one side of the sporangium, a ring of enlarged cells called the **annulus** forms. The annulus functions in opening the sporangium and in releasing the spores. On the opposite side of the sporangium, **lip cells** form that break open, leaving an **aperture**. The aperture increases in size as the annulus cells continue to lose water and be drawn together. Eventually, the annulus cells lose so much water they are no longer held together and the open side of the sporangium snaps forward, propelling the spores out of the sporangium.

10. Using the same slide, locate a single sporangium. Observe the sporangium under low power.

11. Diagram a sporangium below. Label the annulus and spores.

Once spores are released and land in a suitable environment, they germinate into a gametophyte. The gametophyte is a small, heart-shaped, free-living plant approximately the size of a thumbnail. It is composed of a sheet of cells that is attached to the ground by rhizoids. A fern gametophyte is often called a prothallus. Many fern species produce monoecious gametophytes that produce both archegonia and antheridia on the lower surface of the prothallus. Archegonia are usually produced near the notch in the heart-shaped prothallus, whereas antheridia are produced toward the edges of the prothallus.

Archegonia produce nonmotile eggs, whereas antheridia produce flagellated motile sperm cells. The sperm are released from the antheridia into water after rainfall. The sperm are attracted to chemicals released by the archegonia into the water. Sperm detect the chemicals and swim toward the source, eventually swimming into the neck of the archegonium. The egg is housed within the venter. It is fertilized by the sperm, and the zygote remains in the tissues of the archegonium. The zygote begins to develop into a sporophyte. The first leaf of the sporophyte extends upward from the notch in the prothallus.

12. Obtain a prepared slide of a fern prothallus.

13. Observe the specimen under scanning power.

14. Sketch the prothallus below. Label the prothallus, rhizoids, archegonia, and antheridia.

15. Does your specimen contain a new sporophyte leaf? If so, describe its location on the prothallus.

COMPARE AND CONTRAST

1. How are the arrangements of sporangia similar in ferns and horsetails? How are they different?

2. Compare and contrast the functions of the spores and the gametes in the life of a fern.

....

....

....

....

....

3. Fern sporophytes have roots, whereas fern gametophytes have rhizoids. How do these structures relate to the vascular tissues, the sporophyte versus the gametophyte.

....

....

....

....

....

■ Activity 17.4: Whisk Ferns

A small group of highly derived and simplified ferns called the whisk ferns are classified in genus *Psilotum*. The whisk ferns are some of the simplest of all vascular plants. They do not possess true root or true leaves; however, they do have photosynthetic scale-like structures that resemble enations that extend outward from the surface of their stems. The stems are the primary sites of photosynthesis. The body of a whisk fern is relatively simple and consists of a series of upright dichotomously branching stems growing from a horizontal rhizome. Stems are simple and possess a basic vascular arrangement consisting of xylem in the center of the stem surrounded by phloem. The cortex is large. Cells near the outer edge of the cortex are chlorenchyma and are primarily responsible for photosynthesis. Cortex cells in the internal regions of the stem are involved in storage of nutrients. An epidermis covers the cortex. The epidermis contains guard cells and stomata to allow carbon dioxide to enter the stem for photosynthesis. The epidermis also produces a cuticle to limit water loss.

1. Obtain a prepared slide of the cross-section of a *Psilotum* stem.
2. Observe the specimen under scanning power. Find the vascular tissues.
3. Label the xylem, phloem, and cortex on **Figure 17-5**.

The sporangia of *Psilotum* are found in a three-part cluster called a **synangium**. Each synangium consists of three sporangia fused to one another and sharing a common outer wall. The outer wall contains several small **dehiscent cells** that open the wall when they lose sufficient water. Each sporangium produces spores by meiosis. The synangia appear as yellow, rounded clusters resembling buds on the surface of the stems.

4. Obtain a prepared slide of a cross-section of a *Psilotum* synangium.
5. Observe the specimen under scanning power. Locate the three sporangia. Label a spore, sporangium, synangium, and dehiscent cell on **Figure 17-6**.

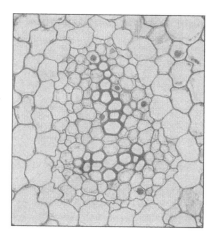

FIGURE 17-5   Cross-section of the vascular structure of *Psilotum* stem (×100). (Courtesy of James D. Mauseth)

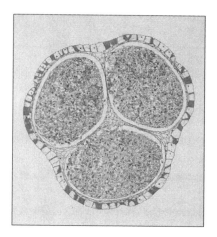

FIGURE 17-6   Sporangia of *Psilotum* (×8). (Courtesy of James D. Mauseth)

COMPARE AND CONTRAST

1. Compare and contrast the arrangement of sporangia in whisk ferns with the arrangement in ferns.

....

....

....

....

....

2. How is the body of a whisk fern similar to that of ferns? How is it different?

3. Describe a similarity and a difference between the whisk ferns and the lycophytes.

Study Guide

- Be able to define the terms in bold.
- Be able to identify all structures related to the reproduction of the seedless vascular plants.
- Be able to differentiate between the key characteristics of the various groups of seedless vascular plants.
- Be able to compare and contrast the divisions of seedless vascular plants.

Conclusions

1. Describe the roles of the sporophyte and the gametophyte in the life cycle of the seedless vascular plants.

2. How are strobili similar to sori and synangia?

3. Which seedless vascular plants possess microphylls?

    ....

    ....

4. Which seedless vascular plants possess megaphylls?

    ....

    ....

5. Compare the function of a sporophyll with that of a megaphyll.

    ....

    ....

    ....

6. Compare and contrast the terms homospory and heterospory. How does the differ-
    ence in spore type affect the gametophyte(s) produced?

    ....

    ....

    ....

7. What is the function of an indusium in ferns? Is there an analogous structure in horse-
    tails? If so, what is it?

    ....

    ....

    ....

8. Define the key parts of a fern frond below.
    a. Rachis

    ....

    b. Petiole

    ....

    c. Lamina

    ....

    d. Pinnae

    ....

9. Lycophytes are sometimes called club mosses. Why are these not considered true mosses?

....

....

....

10. Why are the plants studied in this section larger than the nonvascular plants?

....

....

....

Chapter 18

Gymnosperms

Laboratory Activities

Activity 18.1: Pine Roots

Activity 18.2: Pine Stems

Activity 18.3: Pine Leaves

Activity 18.4: Pine Reproduction

Goals

Following this exercise students should be able to

- Describe the key characteristics of gymnosperms.
- Explain the life cycle of a gymnosperm.
- Understand role of spores and seeds in the life cycle of a gymnosperm.
- Compare and contrast seed cones and pollen cones.

Introduction

Seed plants include the angiosperms and gymnosperms. The evolution of the seed allowed these plants to reproduce without relying on environmental water for fertilization. Seeds function to protect the developing embryo from harsh environmental conditions and are the primary means of dispersal in plants that produce them. **Gymnosperms** produce seeds left exposed to the environment (i.e., they are not enclosed in a carpel). In contrast, angiosperms produce seeds that are enclosed within a carpel and therefore isolated from the outside environment.

Gymnosperm structure is similar to angiosperms in many ways; however, there are a few key differences. The wood of most gymnosperms lacks vessel elements. Instead, tracheids are the main conducting cells. This is the main reason that gymnosperm wood is considered soft wood, unlike the wood of eudicots. Pines, members of genus *Pinus*, are the classic example of a common gymnosperm. Pines produce **resin**, a mixture of organic compounds that plays a role in protecting the plant from microbial diseases. Resin is produced in a series of **resin canals** that extend throughout a pine. The canals are bordered by **glandular parenchyma cells** that secrete the resin into the canals.

Reproduction in gymnosperms follows the same pattern of alternation of heteromorphic generations. Sporophytes are the dominant form. They produce spores by means of meiosis. The spores are not released from the plant. Instead, they remain housed within the tissues of the sporophyte. The spores germinate while contained within the sporophyte tissues to form the gametophytes.

Activities

■ Activity 18.1: Pine Roots

Pine roots have the same basic arrangement of tissues as woody eudicot roots. The bulk of the root is composed of secondary xylem, with secondary phloem produced on the outer side of the wood.

1. Obtain a prepared slide of a cross-section of a *Pinus* root.
2. Observe the specimen using the scanning objective.
3. Locate the secondary xylem. You should see a number of resin canals that appear as clear openings surrounded by small cells. These are found throughout the secondary xylem.
4. Sketch a resin canal below.

COMPARE AND CONTRAST

1. Refer to a cross-section of a *Tilia* woody root. Other than the presence of resin canals, how are *Pinus* roots different from those of *Tilia*?

....

....

....

....

2. Compare the resin canals visible in your specimen in terms of size, shape, and location. Look for overall trends, if any.

....

....

....

....

■ Activity 18.2: Pine Stems

The stems of pine contain tissues arranged in a pattern quite similar to that of woody eudicot stems. The main difference is the presence of resin canals.

1. Obtain a prepared slide of a cross-section of a *Pinus* stem.

2. Observe the specimen under scanning power. Locate the central pith. Are resin canals present?

3. What tissue surrounds the pith?

4. Locate the secondary xylem. Are resin canals present?

5. Locate the secondary phloem. What tissue is found between the secondary phloem and the secondary xylem?

6. Are resin canals present in the secondary phloem?

7. Locate the cortex. Are resin canals present in this region?

8. Label the resin canal, wood, and a tracheid on **Figure 18-1**.

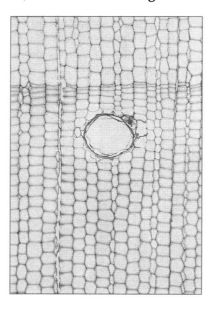

FIGURE 18-1 Pine wood cross-section (×30). (Courtesy of James D. Mauseth)

COMPARE AND CONTRAST

1. Compare the number and size of the resin canals in the secondary xylem with those in the cortex of the stem.

....

....

....

....

....

2. Contrast the number and appearance of resin canals in roots and in stems.

....

....

....

....

....

■ Activity 18.3: Pine Leaves

Leaves of conifers are found in bundles called **fascicles**. Each fascicle may contain two to five leaves. When first produced the leaves are joined closely to one another within the fascicle. As the leaves mature they spread out from the other members of the fascicle to produce a fan-like spread of leaves.

Pine leaves are linear rather than broad and flat. They have several adaptations for water conservation. The stomata are found in a layer of cells just below the **epidermis**, called the **hypodermis**. Based on their location, these are referred to as **sunken stomata**. These leaves also contain an **endodermis** that surrounds the vascular bundles within the leaf **mesophyll**. Inside the endodermis is the vascular bundle. Some species have a single vascular bundle enclosed by the endodermis, whereas others may have several vascular bundles in each leaf. Xylem is found on the side of the vascular bundle nearest the center of the fascicle. Phloem is found toward the outer edge of the vascular bundle. Between the vascular tissues and the endodermis are layers of thin-walled cells called **transfusion tissue**. The mesophyll is not divided into spongy and palisade layers.

1. Obtain a prepared slide of the cross-section of *Pinus* leaves.

2. Observe the specimen with scanning power. How many leaves are present in this fascicle?

....

....

....

3. Are resin canals present in this specimen? If so, how large are they when compared with the size of the vascular tissues inside the endodermis?

....

....

....

4. Label the mesophyll, resin canal, epidermis, and glandular parenchyma cells in **Figure 18-2**.

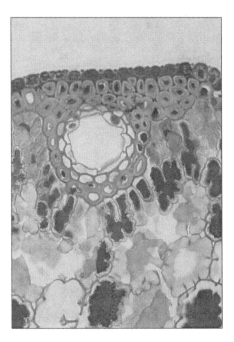

FIGURE 18-2   Cross-section of a pine leaf with resin canal (×160). (Courtesy of James D. Mauseth)

5. Center a vascular bundle in the field of view. Observe the vascular bundle with the low power objective.

6. Label the endodermis, vascular bundle, xylem, phloem, transfusion tissue, and mesophyll on **Figure 18-3**.

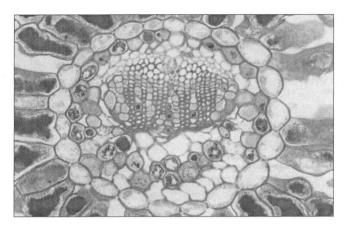

FIGURE 18-3 Cross-section of Douglas fir leaf (×80). (Courtesy of James D. Mauseth)

COMPARE AND CONTRAST

1. Compare and contrast the location of the vascular tissues in pines with those of flowering plants.

2. Where else in a plant can you find an endodermis? How does the endodermis of leaves compare with other endodermis?

■ Activity 18.4: Pine Reproduction

Pines exhibit an alternation of generations much like that of the seedless vascular plants. Spores are contained within sporophyte tissues, whereas seeds are the primary mechanisms for dispersal. Most gymnosperms, pines included, produce two types of cones: seed cones and pollen cones. Often these cones are produced on the same plant, but some species are dioecious, with one plant producing seed cones and one producing pollen cones.

Pollen cones are composed of thin, papery scales attached to a central cone axis. These are simple cones consisting of sporophylls directly attached to the cone axis. No bracts are involved. They are smaller than the seed cones. Each sporophyll supports sporangia. Within each sporangium are numerous microsporocytes that undergo meiosis and produce microspores. These microspores divide by mitosis to form an immature male gametophyte, known as a **pollen grain**.

1. Obtain a prepared slide of the longitudinal section of a pollen cone.

2. Observe the specimen under scanning power. Locate the central cone axis and attached sporophylls. What type of cells can you find in the sporangium?

....

....

....

3. Draw the pollen cone section in the space provided. Label the cone axis, a sporophyll, and a sporangium as identified in **Figure 18-4**.

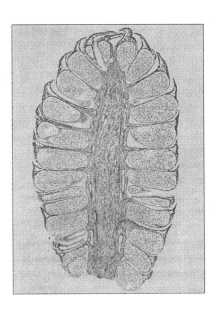

FIGURE 18-4   Longitudinal section of a *Pinus* pollen cone (×10). (© Robert and Linda Mitchell)

The pollen grain of pines is composed of four cells: one **generative cell**, one **tube cell**, and two body cells. The generative cell produces two sperm cells, whereas the tube cell elongates to form the **pollen tube**. The pollen tube is produced once a pollen grain arrives in an ovule. It digests its way through the nucellus and the female gametophyte, into an archegonium, and delivers sperm to the egg for fertilization. These four cells are contained within the central area of the pollen grain. On either side of this region, two **air bladders** are present to provide buoyancy in the air. Pollen grains are released from the pollen tubes into the air. They circulate with air currents, and a small percentage is eventually carried to an ovule of a seed cone.

4. Obtain a prepared slide of *Pinus* pollen grains.

5. Diagram a pollen grain below. Label the **air bladders**.

Gymnosperms and other seed plants produce **ovules**. Ovules are bound by a protective wall called the **integument**. The ovules house the megasporangium, the megaspore, the female gametophyte, and ultimately the eggs. Once an egg is fertilized, the integument hardens to form the **seed coat**. The embryo develops within the ovule, and the entire structure is now called a **seed**. The seeds are released when they are mature and germinate when they reach soil in a suitable habitat.

The ovule has an opening in one end called the **micropyle**. This is the opening through which pollen grains enter in the process of pollination. Those pollen grains reset in a small open space called the **pollen chamber** that is found just within the micropyle. The pollen grains rest on the remnants of the megasporangium, called the **nucellus**.

Ovules are produced on the surfaces of **ovuliferous scales**, which are composed of bracts that branch off the central cone axis. A bract supports a bud that gives rise to a sporophyll. The bracts fuse with the sporophylls to form an ovuliferous scale. Numerous ovuliferous scales are attached to one central cone axis, forming a complex cone, so considered because the bracts contain additional buds. Ovules are found in pairs at the base of each ovuliferous scale where they attach to the axis. Collectively, these form the female pine cone, or **seed cone**. Seed cones are large and woody and are the pine cones with which most people are familiar.

The ovules, when first produced, contain a **megasporangium** with **megasporocyte** cells that divide by meiosis to produce four haploid cells. Three of those degenerate, leaving one large, haploid **megaspore**. This megaspore undergoes multiple rounds of mitosis, producing a **female gametophyte** consisting of several thousand cells. The female gametophyte produces at least two **archegonia**. Each archegonium produces a single **egg**.

6. Obtain a prepared slide of a longitudinal section of a *Pinus* seed cone.

7. Observe the specimen on scanning power. Note the ovuliferous scales found attached to the central cone axis.

8. Label an ovuliferous scale, the cone axis, and an ovule on **Figure 18-5**.

9. Obtain a prepared slide of a longitudinal section of a *Pinus* ovule.

10. Observe the specimen under low power. Locate the female gametophyte within the ovule. Center it in your field of view. High power may be used to view the specimen if needed.

11. Label the integument, micropyle, female gametophyte, pollen chamber, and ovuliferous scale in **Figure 18-6**.

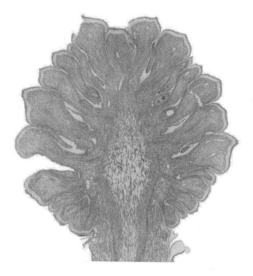

FIGURE 18-5   Longitudinal section of a *Pinus* seed cone (×2). (© Thomas Mounsey/ShutterStock, Inc.)

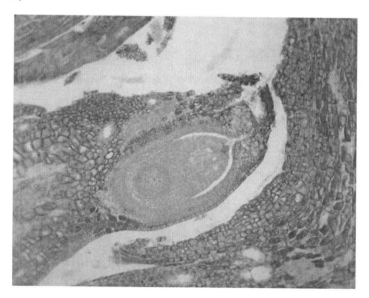

FIGURE 18-6   Longitudinal section of *Pinus* ovule. (© Jubal Harshaw/ShutterStock, Inc.)

Once the egg is fertilized, it forms a zygote that subsequently divides by mitosis to form an **embryo**. The embryo cells form the first root, the **radicle**, and shoot, and the **epicotyl**. The epicotyl is surrounded by **cotyledons**, often called seed leaves, that provide some stored nutrients to the embryo during germination. This embryo is contained within the remaining tissues of the female gametophyte.

12. Obtain a prepared slide of the longitudinal section of a *Pinus* seed.

13. Observe the specimen under scanning power.

14. Label the seed coat, embryo, radicle, cotyledons, female gametophyte, and epicotyl in **Figure 18-7**.

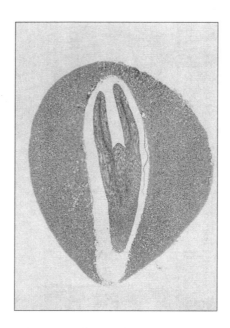

FIGURE 18-7   Longitudinal section of a conifer seed (×25). (© Phototake/Alamy Images)

COMPARE AND CONTRAST

1. How are the structures of seed cones and pollen cones similar? How can they be distinguished from one another?

2. Compare the process of pollination with the process of fertilization.

3. What is the role of the ovule in the reproduction of gymnosperms? What, if any, are equivalent structures in seedless vascular plants? Explain your answer.

4. Explain how a pollen grain differs from a microspore.

Study Guide

- Be able to define the terms in bold.
- Be able to describe the life cycle of a pine.
- Be able to differentiate the roles of the sporophyte and the gametophyte in the life cycle of a gymnosperm.
- Be able to describe the structure of an ovule and relate it to the structure of a seed.

Conclusions

1. Why are pines considered to be a source of softwood instead of hardwood, such as that produced in oaks?

2. Is the gametophyte independent of or dependent on the sporophyte? Explain your answer.

3. Compare and contrast pollen cones and seed cones.

4. How can you explain the plentiful amount of resin canals present throughout a pine?

5. Compare the roles of spores and seeds in the life of a gymnosperm.

6. Label each of the following as haploid or diploid.
    a. Ovuliferous scale

    b. Female gametophyte

    c. Nucellus

    d. Integument

    e. Archegonium

    f. Egg

7. Explain how pollination differs from fertilization. Which happens first?

8. Describe two adaptations of a pine leaf against excess water loss through transpiration.

Chapter 19

Angiosperms I: Flowers

© Cristi Matei/ShutterStock, Inc.

Laboratory Activities

Activity 19.1: Tulip (*Tulipa* sp.) or Gladiolus (*Gladiolus* sp.)

Activity 19.2: Lily (*Lilium* sp.)

Activity 19.3: Geranium (*Pelargonium*)

Activity 19.4: Rose (*Rosa* sp.)

Activity 19.5: Petunia (*Petunia* sp.)

Activity 19.6: Snapdragon (*Antirrhinum majus*)

Activity 19.7: Daisy or Chrysanthemum (*Chrysanthemum* sp.)

Activity 19.8: Ovule Development

Activity 19.9: Pollen Development

Goals

Following this exercise students should be able to

- Describe the parts of a flower.

- Understand the role of a flower in the life cycle of an angiosperm.

- Contrast reproduction in angiosperms with that of gymnosperms.

Introduction

Flowers are reproductive structures produced by only one class of plants, the Magnoliopsida. The structures of flowers vary widely; however, there are four basic **whorls** (layers) in a flower, each serving a different function. The whorls are attached to the apical portion of a modified stem called the **pedicel**. The upper end of the pedicel supports a receptacle to which the flower whorls attach.

The outermost whorl is the **calyx**, comprised of **sepals** that are usually green and leaf-like. The primary function of the calyx is to protect the bud before it opens to reveal the flower. The **corolla** is the second whorl and includes the **petals**. Petals are typically brightly colored and function in recruiting pollinators. The calyx and corolla together form the **perianth**. The stamens comprise the third layer. Each **stamen** has two parts: the anther, which produces pollen grains, and the filament, which supports it. Collectively, the stamens are referred to as the **androecium**. The innermost whorl is made up of the carpels. Carpels are modified leaves that house sporangia. In these plants the carpels also eventually house the egg. The carpels form a structure called the **pistil**. Pistils have three parts: the **stigma**, which receives pollen; the **style**, which elevates the stigma; and the **ovary**, which houses the egg.

Each pistil may be composed of one carpel, in which case it is a simple pistil, or it may be made up of multiple fused carpels, in which case the pistil is compound. Likewise, the terms simple and compound can refer to the ovaries composed of one (simple) or multiple (compound) carpels. The carpels are collectively referred to as the **gynoecium**. The androecium and gynoecium are the reproductive parts of the flower. The ovary contains multiple ovules. Each ovule is found within a **locule**, a small cavity within the ovary. Each ovule is attached to a specific area of the ovary wall called the **placenta**.

Some flowers are composed of all four whorls and called **complete flowers**. **Incomplete flowers** are missing one or more whorls. Incomplete flowers may be adapted for wind pollination where petals and sepals prevent wind from carrying pollen effectively. They may result as a specialized type of reproduction to increase the rate of cross-pollination. **Imperfect flowers** contain either an androecium or a gynoecium but not both. Hence, the pollen from one flower must pollinate the stigma of another. **Perfect flowers** contain both an androecium and a gynoecium and may self-pollinate or cross-pollinate.

Pollination is completed as the pollen from an anther reaches the stigma of a pistil. Once the pollen grain rests on the stigma, chemical signals cause it to germinate. The pollen grain produces a long, thin pollen tube that grows down the stigma into the ovary, digesting the stigma tissues as it proceeds downward until it reaches an ovule within the ovary.

The female gametophyte is much reduced in the angiosperms relative to gymnosperms or other plant groups. Within each ovule a female gametophyte forms from the megaspore. Each female gametophyte consists of seven cells: three antipodal cells, one egg cell, two synergid cells, and a central cell with two nuclei. There are no archegonia. The egg and two synergid cells are found near the micropyle, whereas the three antipodal cells are located at the opposite end of the ovule. The pollen tube delivers two sperm cells to the female gametophyte. One sperm fertilizes the egg to form a zygote. The other sperm combines with the two nuclei of the central cell to form a triploid **endosperm** nucleus. This process is called **double fertilization**.

Remember, angiosperms are seed plants like gymnosperms. Even though their reproductive structures differ, their life cycles are similar. The sporophyte is the dominant generation, and the gametophytes depend on the sporophyte. Like gymnosperms, angiosperms produce an ovule. Ovules in both types of plants serve the same function and have similar structures. The ovule forms the seed once the egg within is fertilized.

The earliest flowers exhibited radial symmetry, where the flower could be divided into mirror-image halves by multiple vertical planes. These primitive flowers also had large numbers of floral parts. As flowers evolved and become more advanced, the number of floral parts became reduced and bilateral symmetry developed in some lineages. Bilaterally symmetrical flowers can only be divided into mirror-image halves by one plane, creating left and right halves. Advanced flowers tend to have a higher degree of fusion between the floral parts as well.

In **hypogynous** flowers the petals and sepals attach below the gynoecium. The ovary in such a flower is said to be **superior**. Conversely, in **epigynous** flowers the perianth is attached above the gynoecium. The gynoecium in these flowers is embedded within the receptacle and the ovaries are **inferior**. In some advanced flowers the receptacle fuses to the gynoecium. Some flowers are neither hypogynous nor epigynous. Instead, these have ovaries that are partially embedded within the receptacle. The term **perigynous** refers to sepals and petals that attach along the side of the ovary. The ovary can also be described as **half-inferior**.

Monocots and eudicots share the same basic floral structures; however, the number of floral parts can be used to differentiate between monocots and eudicots. Monocots have flowers that are composed of three, or multiples of three, parts. Eudicots, on the other hand, are composed of flower parts arranged in multiples of four or five. Thus, by counting sepals, petals, stamens, or pistils you can determine if the flower you are studying is from a monocot or a eudicot.

Activities

■ Activity 19.1: Tulip (*Tulipa* sp.) or Gladiolus (*Gladiolus* sp.)

1. Obtain a tulip or gladiolus flower. How many sepals are present? How many petals are present?

2. Describe the symmetry of this flower.

3. Cut the flower longitudinally. How many stamens are present?

4. Remove a stamen from the flower. Diagram it below and label the anther and filament.

5. How many pistils are present?

6. Isolate the pistil from the rest of the flower. Observe the stigma under a magnifying glass. How many lobes are present?

7. Is this an example of a superior or inferior ovary?

8. Diagram an intact pistil below. Label the stigma, style, and ovary.

9. Cut the specimen to make a cross-section of the ovary. Observe it under a magnifying glass. Diagram the cross-section of the ovary below. Label an ovule, locule, and carpel.

10. Is this a simple or compound pistil?

COMPARE AND CONTRAST

1. How are carpels and pistils similar? How can you tell them apart?

2. Compare the structure of the stigma and style with their functions.

■ Activity 19.2: Lily (*Lilium* sp.)

1. Obtain a lily flower. Describe its symmetry.

2. How many sepals are present? Are they fused or free?

3. How many petals are present? Are they fused or free? What color are they?

4. Cut the lily longitudinally. How many stamens and pistils are present?

5. Are the ovaries superior or inferior?

6. Diagram the stigma below. Is this representative of a simple or compound pistil?

7. How many carpels are present?

8. Make a cross-section of the ovary. Diagram it below and label an ovule, locule, and carpel.

COMPARE AND CONTRAST

1. Compare and contrast the structure of a stamen with that of a pistil.

2. How are lilies similar to tulips or gladiolus?

Activity 19.3: Geranium (*Pelargonium*)

1. Obtain a flower from a florist's geranium. Describe the perianth.

2. Remove the stamens from the flower. Diagram a stamen below and label the anther and filament.

3. Is this flower epigynous or hypogynous?

4. How many pistils are present? Diagram an intact pistil below.

5. Cut the ovary across the middle. Diagram the ovary cross-section below. How many carpels are present?

6. Is this flower from a monocot or a eudicot?

....

....

....

COMPARE AND CONTRAST

1. Compare and contrast the structure of a geranium with that of a lily. Include symmetry, number of floral structures, and any other relevant characteristics.

....

....

....

....

....

2. How are the structures of sepals and petals similar?

■ Activity 19.4: Rose (*Rosa* sp.)

1. Obtain a rose flower. What type of symmetry does it exhibit?

2. How many sepals are present? Sketch one below.

3. Remove the petals from the rose. Notice they become increasingly smaller as you move toward the center of the flower. Align them from largest to smallest. Diagram a large outer petal and a small inner petal below.

4. The innermost petals surround a layer of rudimentary anthers that are characterized by poorly developed anthers and a stamen that is somewhat petal-like in appearance. Compare these rudimentary anthers with the well-developed anthers on the stamens. Diagram both below.

5. Make a longitudinal cut through the center of the rose gynoecium and receptacle. Diagram the longitudinal section below. Label a stamen, pistil, and the receptacle.

6. Is this an epigynous or hypogynous flower? Is there another term that would describe it better?

....

....

....

7. How many pistils are present? Remove one and diagram it below. Label the stigma, style, and ovary.

8. How many carpels for a single pistil? Make a cross-section of the ovary if necessary to answer this question.

9. Is this flower primitive or advanced? Explain your answer.

COMPARE AND CONTRAST

1. How are the sepals of a rose similar to rose leaves? How are they different?

2. How are the rudimentary anthers similar to the well-developed anthers? How do they differ?

■ Activity 19.5: Petunia (*Petunia* sp.)

1. Obtain a flower of petunia. Observe its external features. What type of symmetry does the petunia flower possess?

2. How many sepals are present? Are they fused or free?

....

....

....

3. How many petals are present? Are they fused or free?

....

....

....

4. Diagram a side view of an intact petunia flower. Label the calyx and corolla.

5. Remove the perianth. Sketch the intact androecium and gynoecium in the space below.

6. How many stamens are present? How many pistils are present?

....

....

....

7. Is this flower primitive or advanced?

COMPARE AND CONTRAST

1. Compare and contrast the external structure of a petunia with that of a rose and a geranium.

2. How do the androecium and gynoecium compare between petunias and roses?

■ Activity 19.6: Snapdragon (*Antirrhinum majus*)

1. Obtain a snapdragon flower. Describe its symmetry.

2. How many sepals are present? Are they fused or free?

3. Notice the petals are fused together, forming a corolla tube. The tube is divided into upper and lower parts. How many petals form each part?

---

4. This flower is modified for pollination by certain types of insects, such as bees, that are able to push through the opening in the petals. Gently pull the petals apart. What do you see?

---

5. Remove the corolla tube. Notice the stamens are fused to the petals. How many stamens and pistils are present? Describe their arrangement.

---

6. Make a cross-section of the ovary using a razor blade. Examine it with a hand lens or dissecting microscope. How many carpels are present? Is this ovary simple or compound?

---

7. Diagram the cross-section of the snapdragon ovary. Label the ovary wall, carpels, ovules, and locules.

8. Is this a representative of primitive or advanced flowers? Why?

COMPARE AND CONTRAST

1. Both petunias and snapdragons have fused floral parts. Compare the fused parts in both types of flowers.

2. How are the corolla tubes similar in petunias and snapdragons? How are they different?

■ Activity 19.7: Daisy or Chrysanthemum (*Chrysanthemum* sp.)

1. Obtain a daisy or chrysanthemum. Observe its external features. How many sepal and petals are present?
2. You should have counted zero sepals and five petals. If you counted more than that, you made a common error. These "flowers" are actually clusters of many small flowers called an **inflorescence**. This particular inflorescence is called a head that has a short central axis. The flowers produced by this axis therefore remain close to one another. These flowers, along with sunflowers, black-eyed Susan, coneflowers, and many others are in a family called **composites**. The daisy type of head contains two types of flowers: **ray flowers** around the outer edge and **disk flowers** in the center of the head. The green structures below the head are bracts, leaves that support reproductive structures. How are bracts similar to sepals?

3. Slice the head in half longitudinally. Examine the cut edge. The base of each flower contains the ovary, a small oval-shaped structure. Remove an intact ray flower, including the basal ovary. What is the symmetry of this ray flower?

4. Notice that the ray flower has no sepals. Describe its five petals. Are they fused or free?

5. Is this a hypogynous or epigynous flower?

6. These ray flowers do not contain stamens. Diagram the intact pistil below. Label the stigma, style, and ovary. Although pistils are present, these ray flowers are typically sterile.

7. Locate the disk flowers in the cut section of the inflorescence. Notice disk flowers are in several stages of development. Compare the locations of the most developed and least developed disk flowers.

8. Remove an intact, mature disk flower. Observe it with a hand lens or dissecting microscope. Describe the symmetry of this flower.

9. The petals of the disk flower are fused, forming a corolla tube. How many petals are present?

10. Split the corolla tube in half longitudinally. Notice the stamens within the corolla tube. How many stamens are present? Are they fused or free?

11. Find the single pistil. Is this a simple or compound pistil? What evidence do you have to support your answer?

COMPARE AND CONTRAST

1. Compare the symmetry and floral structure of a ray flower and a disk flower.

2. Snapdragons are another type of inflorescence with a long axis. Compare the arrangement of flowers in the snapdragon inflorescence with those in the *Chrysanthemum* inflorescence.

■ Activity 19.8: Ovule Development

The ovules contained in the ovary of a flower house a megasporangium. Each megasporangium contains a megasporocyte that divides by meiosis to produce four haploid cells. One of these forms the megaspore and the other three degenerate. The megaspore divides by mitosis to produce the female gametophyte. Each female gametophyte contains a single egg.

1. Obtain a prepared slide of the cross-section of a *Lilium* ovary.
2. Observe the specimen under scanning power. Locate the ovules and center them in the field of view.
3. Observe the ovules under low power. The ovule is attached to the locule by a funiculus, an extension of the ovary tissues that contains both xylem and phloem.
4. Label the ovule, megasporangium, and funiculus in **Figure 19-1**.

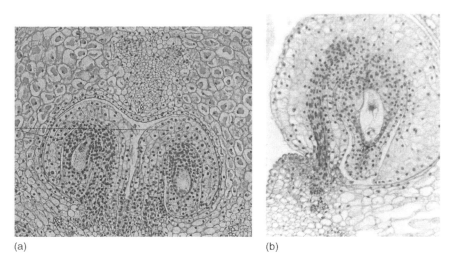

(a)   (b)

FIGURE 19-1   Cross-section of *Lilium* ovary (a, ×50; b, ×120). (Courtesy of James D. Mauseth)

■ Activity 19.9: Pollen Development

The anthers are the sites of pollen production in a flower. Anthers contain numerous microsporocytes that divide by meiosis to produce a tetrad of haploid cells. This occurs within circular areas bordered by nurse cells. They later separate and take on their mature shapes, at which time they are called pollen.

1. Obtain a prepared slide of the cross-section of *Lilium* anther.
2. Observe the anthers under scanning power. Diagram the general structure of the anther cross-section.

3. Locate one area of microspore formation and observe it under low power. High power may be necessary to see the spores in detail.

4. Label the nurse cells and developing microspores on **Figure 19-2**.

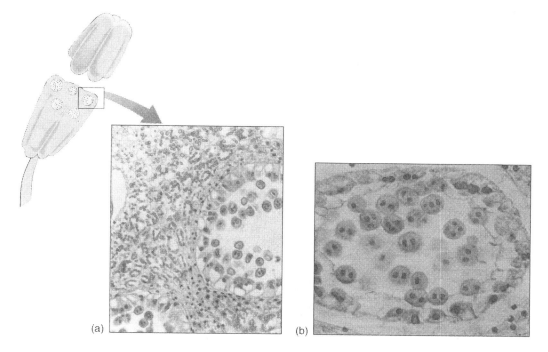

(a)

(b)

FIGURE 19-2   Cross-section of *Lilium* anther (a, ×200; b, ×500). (Courtesy of James D. Mauseth)

5. The anther opens by dehiscence at weak areas in the anther wall. This releases the pollen (immature male gametophytes). Can you see pollen being released in your specimen? If so, sketch one below.

Study Guide

- Be able to define the terms in bold.
- Be able to describe the reproductive strategy of angiosperms.
- Be able to identify the parts of a flower and describe the role of each in reproduction.
- Be familiar with trends in floral evolution.

Conclusions

1. Explain the difference between hypogynous and epigynous flowers.

   ....

   ....

   ....

2. Sepals and petals are modified leaves. Describe the characteristics of each that are shared with leaves.

   ....

   ....

   ....

3. A flower consists of sepals, stamens, and a pistil. Is it complete or incomplete? Is it perfect or imperfect?

   ....

   ....

   ....

4. Describe the role of the stigma in pollination.

   ....

   ....

   ....

5. Explain the relationship between carpels and pistils.

   ....

   ....

   ....

6. Define the four whorls of a typical flower.

   ....

   ....

   ....

7. Label each of the following structures as haploid or diploid.
   a. Megasporocyte

   ....

   b. Microspore

   ....

c. Pollen grain

....

d. Female gametophyte

....

e. Egg

....

f. Anther

....

8. Describe the perianth of a rose.

....

....

....

9. Compare and contrast bilateral and radial symmetry.

....

....

....

10. Contrast the characteristics of primitive flowers with those of advanced flowers.

....

....

....

Chapter 20

Angiosperms II: Fruits

Laboratory Activities

Activity 20.1: Seeds

Activity 20.2: Fleshy Fruits

Activity 20.3: Dry Dehiscent Fruits

Activity 20.4: Dry Indehiscent Fruits

Goals

Following this exercise students should be able to

- Identify the tissues comprising a fruit.
- Identify various types of fruit.
- Understand the role of fruit in the life of a flowering plant.

Introduction

Fruits are only produced by angiosperms. Once an egg is fertilized, the ovule goes through a number of changes to become a **seed**. The seeds are enclosed within the tissues of the ovary, which ripen to form a fruit. A single fruit may contain a single seed or multiple seeds. Fruits sometimes include other tissues beyond those of the ovary, most often part of the receptacle of an inferior ovary. Occasionally, sepals, petals, or stamens can also be incorporated into a fruit.

The tissues comprising a fruit are called the **pericarp**. This is a generalized term and does not reflect the origin of those tissues. True fruits are those that contain tissues derived from the ovary alone. Accessory fruits contain tissues from the ovary as well as from nonovarian sources, such as the receptacle. The innermost part of an accessory fruit is composed of carpel-derived tissues.

Some true fruits develop from the tissues of a single ovary (simple pistil) or multiple fused ovaries of one flower (compound pistil). These are **simple fruits** and are the most plentiful type of fruit. When a fruit develops from separate carpels (i.e., separate pistils) of one gynoecium, an **aggregate fruit** forms. A common example of an aggregate fruit is a blackberry. **Multiple fruits** form when the pistils from multiple flowers in an inflorescence fuse. Figs and pineapples are common examples of multiple fruits.

Fruits are also classified by their physical characteristics. **Fleshy fruits** contain soft tissues that are normally eaten by animals. **Dry fruits** are not normally eaten because they are made from harder tissues and are generally not palatable. Dry fruits can be further grouped into dehiscent and indehiscent fruits. **Dehiscent fruits** split open when mature to release seeds, whereas **indehiscent fruits** retain the seed within the fruit and do not split open naturally.

Activities

■ Activity 20.1: Seeds

Seeds are the reproduction structures that protect the embryo. In flowering plants they are the primary mechanism for dispersal of the plant. Seeds begin to form at fertilization. Remember, flowering plants go through double fertilization. One sperm fertilizes the egg, forming a zygote that subsequently develops into the embryo. The other sperm fertilizes the central cell nuclei to form the endosperm nucleus that develops into a tissue by the same name. The endosperm is triploid and functions to provide nutrients for the developing embryo.

Embryos contain at least one **cotyledon**. Cotyledons are commonly referred to as seed leaves. In eudicots (eudicotyledons) and basal angiosperms, each embryo possesses two cotyledons, whereas monocot embryos possess only one cotyledon. The cotyledons of eudicots become enlarged and thicken and absorb most of the nutrients from the endosperm. The intact seed is mostly filled by these cotyledons. In monocots the cotyledon remains thin and the endosperm remains prevalent in the seed. Thus, most nutrients are stored within the endosperm.

The embryo also develops a shoot with stem and leaves. The embryonic leaves are called the **plumule**. The embryonic stem is divided into two regions, the **hypocotyl**, which is below the cotyledons, and the **epicotyl**, which is above the cotyledons. In most seeds the hypocotyl is the only stem visible. The epicotyl elongates when the seed begins to germinate. The embryo also produces an embryonic root, the **radicle**, which is typically the first structure to emerge from the seed coat.

1. Obtain a garden bean that has been soaking in water. The beans may have begun to sprout, in which case the seed coat will often be loose and easy to remove.
2. Observe the external features of the seed. The seed coat, called the **testa**, forms the outer protective barrier around the seed. Two marks are present on the testa. The first is the **hilum**, which marks the location where the seed was attached to the placenta. It is fairly large and generally light in color. The **micropyle** is the original opening of the ovule. It is tiny and may need to be magnified to be visible. Usually, the micropyle is near the hilum.
3. Diagram the garden bean to show the hilum and micropyle. Label the hilum, micropyle, and seed coat.

4. Remove the seed coat. Observe the embryo. Locate the plumule, radicle, hypocotyl, and cotyledons. The radicle is located at one end of an axis of forming root and stem tissues, with the hypocotyl and plumule at the opposite end.

5. Label the cotyledon, plumule, radicle, and hypocotyl on the embryo in **Figure 20-1**.

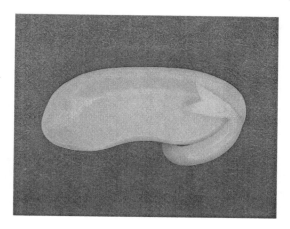

FIGURE 20-1   Bean seed that has begun germinating. (Courtesy of James D. Mauseth)

COMPARE AND CONTRAST

1. The garden seed is mostly filled by cotyledons. How are these different from the leaves of the plumule?

....

....

....

....

....

2. Compare the structure, location, and function of the hypocotyl and radicle.

....

....

....

....

....

▪ Activity 20.2: Fleshy Fruits

Drupes are the simplest of the fleshy fruits. They are commonly called stone fruits because the seeds are encased within a stony pit. The tissues of a drupe are all ovarian; thus, it is a simple fruit. A simple fruit contains three layers of tissues that are easy to identify in a drupe. The outer "skin" of a drupe is called the exocarp. Typically, it surrounds a thicker and usually somewhat soft mesocarp. The endocarp in drupes forms the wall of the pit. It is made up of sclerenchyma and is usually quite hard. The seed is housed within the endocarp. The term "pericarp" refers to all three of these layers in a drupe.

1. Obtain a cherry or other drupe provided. Make a longitudinal cut through the center of the cherry.
2. Diagram the cut side of the longitudinal section below. Label the exocarp, endocarp, and mesocarp.

Berries are simple fruits that have fleshy endocarps. They usually contain multiple seeds. Some common examples of berries are grapes and tomatoes. Many of the fruits we commonly call berries, like strawberries and raspberries, are not botanical berries. The exocarp, mesocarp, and endocarp are similar to those seen in drupes; however, the endocarp is soft and flexible. It may appear quite similar to the mesocarp but is usually discernable by a slight change in color. The endocarp is the innermost region of the fruit.

3. Obtain a grape. Make a longitudinal cut through the middle of the grape. Some grapes and other common fruits are grown as seedless varieties. These grapes have been bred to have a genetic error that prevents seeds from fully developing.
4. How many seeds do you see?

5. Diagram the longitudinal section of a grape below. Label the exocarp, mesocarp, and endocarp.

   Tomatoes are berries that are typically larger than grapes; however, unlike grapes they contain numerous edible seeds. The endocarp of tomatoes forms the lining of the locules. Locules are simply spaces that house the seeds. In tomatoes the locules are large and contain numerous seeds surrounded by a gelatinous substance. A seed is attached to the fruit wall at a region called placenta.

6. Obtain a tomato. Observe the outer structure of the tomato. You should see sepals at the end where the tomato was attached to the stem. At the opposite end of the tomato you should find a small, prickly structure extending out of the exocarp. This is the remnant of the stigma and style.

7. Based on this information, was the tomato flower hypogynous or epigynous?

8. Slice a tomato in half midway between the sepals and the stigma remnant.

9. Diagram the cut side of this tomato cross-section below. Label the exocarp, endocarp, mesocarp, locule, and seeds.

   Pomes are fleshy accessory fruits. These fruits contain tissues derived from the ovary and from the receptacle. The fleshy part of the fruit that is typically consumed is derived from the ovary. Pears and apples are the best-known examples.

   The outermost layers of tissue in an apple, including the fleshy part we eat, are derived from the receptacle. If you look closely at an apple that has been cut longitudinally, you should be able to see the exocarp, which looks like a fine line bordering the apple's core. The endocarp appears as a shiny border on the inner surface of the space (locule) containing the seeds. The mesocarp is found between the exocarp and endocarp.

10. Obtain an apple. Make a longitudinal cut through the center of the apple.

11. Diagram the cut side of the apple below. Label the seeds, exocarp, endocarp, mesocarp, receptacle tissue, sepals, and locules.

12. Based on the location of the sepals relative to the fruit, was this ovary superior or inferior? Was the ovary hypogynous or epigynous?

13. Observe the cross-section of an apple on display. How many carpels were present in the ovary?

14. Diagram the cross-section of the apple below. Label the receptacle tissue, endocarp, exocarp, mesocarp, locule, and seed.

Aggregate fruits are formed when the ovaries of multiple pistils within a single flower fuse together. The result is a fruit with subdivisions representing the contribution of each ovary. Each subdivision is called a **druplet** and contains a single seed.

15. Obtain a blackberry and examine its external features. Notice the small rounded druplets of the fruit. Each druplet is attached to the receptacle that is retained in the center of the fruit.

16. Slice the blackberry longitudinally through the middle of the fruit. Diagram the longitudinal section below and label a druplet and the receptacle.

Strawberries are another example of aggregate fruits that contain some accessory tissues, such as the receptacle. They are sometimes referred to as aggregate-accessory fruits for this reason.

17. Obtain a strawberry and examine its external features. Notice the small oval "seeds" on the outer surface of the fruit. These are actually small dry fruits called achenes.

Each achene contains a seed. The calyx is still attached to the strawberry. If removed, it should reveal numerous small stamens still attached to the fruit.

18. Cut the strawberry in half longitudinally. Observe the cut surface. The receptacle forms the large, fleshy part of the strawberry. Within the receptacle, you should see some light pink or white lines extending outward toward the achenes on the surface. These are vascular bundles.

19. Diagram the longitudinal section of the strawberry below. Label the achenes, vascular bundles, and receptacle.

Multiple fruits are composed of flowers within an inflorescence fusing together during fruit formation. Pineapples are excellent examples of multiple fruits.

20. Observe the longitudinal sections of pineapples on display. Each subdivision seen on the outer surface is the result of a single flower.

21. Look at the cut side of the pineapple. The receptacle forms a central core of fleshy tissue. Each flower is attached to the receptacle and extends outward from the receptacle.

22. Diagram the cut side of the pineapple below. Label the receptacle and a seed. Circle the area contributed by a single flower.

COMPARE AND CONTRAST

1. Compare the structures of simple, aggregate, and multiple fruits.

2. How is the pericarp of a drupe similar to that of a berry? How is it different?

3. Contrast the structures of strawberries and apples, both a type of accessory fruit.

_____

_____

_____

_____

_____

▪ Activity 20.3: Dry Dehiscent Fruits

Dehiscent fruits split open at maturity to release their seeds. A capsule is composed of more than one carpel fused together. Capsules split along the lines joining the carpels and thus can split open in a variety of patterns. Follicles and legumes are elongated dry fruits. Follicles open along one seam, whereas legumes open along two seams.

1. Observe a peanut in the shell. The outer "shell" of the peanut is the pericarp.
2. Open the pericarp. Notice it naturally splits along two seams. This reveals the seeds within. The seeds are the parts that are generally eaten.
3. Diagram the open pericarp and seeds below and label each.

4. Observe the other dehiscent fruits on display and follow your instructor's directions concerning them.

COMPARE AND CONTRAST

1. How does the pericarp of a legume compare with the pericarp of a drupe?

_____

_____

_____

_____

_____

2. If you were to cut open a raw peanut seed, what would you expect to find?

....

....

....

....

■ Activity 20.4: Dry Indehiscent Fruits

Indehiscent fruits do not open when mature. **Samaras** are characterized by large, wing-like extensions that branch off of the center of the fruit. Within the fruit is a single seed. Maples produce samaras. True **nuts** have a hard, stony pericarp and contain a single seed. The seeds of nuts are not fused to the pericarp. Acorns and hazelnuts are true nuts. Many things we commonly call nuts, such as walnuts, almonds, and peanuts, are not nuts at all. Walnuts and almonds are the pits of drupes, whereas peanuts are seeds. **Caryopses**, better known as **grains**, are indehiscent fruits that contain a single seed that is attached to the pericarp so tightly the pericarp cannot be removed from the seed. Corn and wheat are common grains. **Achenes** are similar to a grain; however, the seed is not attached to the pericarp.

1. Obtain a corn grain that has been soaking in water. Examine its external features. At one end of the grain is a **fruit scar** where the grain was attached to the kernel. The pericarp is thin and transparent. The color is generated by the seed coat within.

2. On one of the flat sides of the corn kernel you should see a white, oblong area. This marks the location of the **embryo** within the seed. Most of this white area is the single **cotyledon**, sometimes called the **scutellum**. In the midst of the cotyledon you should see a small ridge. The upper part, away from the fruit scar, represents the embryonic leaf called the **plumule**. The lower part is the **hypocotyl**, an embryonic stem. The rest of the kernel is filled with **endosperm**, which provides nutrients for the embryo and appears yellow.

3. Diagram the side view of a corn kernel (embryo facing up) below. Label the hypocotyl, plumule, endosperm, and cotyledon.

4. Make a longitudinal section through the embryo. In this view the endosperm occupies approximately three-fourths of the internal space within the grain. In one corner you will find the embryo.

5. Diagram the longitudinal section of a corn grain below. Label the scutellum (cotyledon), plumule, pericarp, endosperm, and radicle. The radicle is an embryonic root found at the opposite end of the linear structure that includes the plumule. The scutellum forms the top border of the embryo and covers the plumule.

6. Observe the other indehiscent fruits on display and follow your instructor's directions concerning them.

COMPARE AND CONTRAST

1. How are achenes and grains similar? How are they different?

....

....

....

....

....

2. Compare the structure of the corn embryo with the structure of the garden bean embryo.

....

....

....

....

....

Study Guide

- Be able to define the terms in bold.
- Be able to identify the various types of fruits seen today.
- Be able to explain the roles of fruits and seeds in angiosperm reproduction.
- Understand the importance of fruit structure with reference to mechanism of dispersal.

Conclusions

1. Is an apple a monocot or a dicot? How can you determine this based on fruit alone?

2. Describe the exocarp, endocarp, and mesocarp of a drupe.

3. Why are cotyledons important in a eudicot seed?

4. Compare the structure of a seed with the structure of an ovule.

5. What is the primary purpose of a fruit?

6. The ovule develops into the seed. What structure(s) forms the fruit?

7. How does the cotyledon of corn compare with the cotyledon of a garden bean?

8. What is the function of endosperm?

....

....

....

9. How are berries different from drupes?

....

....

....

10. Compare and contrast dry dehiscent fruits and dry indehiscent fruits.

....

....

....

Chapter 21

© Cristi Matei/ShutterStock, Inc.

Fungi

Laboratory Activities

Activity 21.1: Slime Molds

Activity 21.2: Zygomycetes

Activity 21.3: Ascomycetes

Activity 21.4: Basidiomycetes

Activity 21.5: Imperfect Fungi

Activity 21.6: Lichens

Activity 21.7: Mycorrhizae

Activity 21.8: Rusts and Smuts

Goals

Following this exercise students should be able to

- Differentiate between the major groups of fungi.
- Identify reproductive structures in the fungi.
- Understand a general fungal life cycle.

Introduction

Fungi are eukaryotic heterotrophs that obtain their food by absorption. No fungi are autotrophic. None has chlorophyll. Some are **saprotrophs** that feed on dead, decaying organic material. Others are **biotrophs** that feed on the tissues of living organisms. Still others are **necrotrophs**, pathogens that feed on the tissues of a host organism to the point that the host dies. Despite diversity in form and habitat, they are all classified within **Kingdom Fungi**.

At one time fungi were classified with plants; however, they were placed in their own kingdom due to the many distinctions between plants and fungi. Although some forms of fungal growth superficially resemble plants, many forms do not. Fungi are primarily multicellular filamentous organisms; however, some unicellular species exist.

Fungal cells are bordered by a cell wall that functions similarly to the plant cell wall. The wall is composed primarily of a structural polysaccharide called **chitin**. Chitin is not found in plants; however, it is the chief component of arthropod exoskeletons.

The typical fungal body is composed of filamentous strands of cells called **hyphae**. A hypha may be **septate** and exhibit individual cells separated by cross walls. Conversely, some hyphae are **coenocytic** and lack cross-walls between cells. These hyphae are essentially long multinucleate cells. The fungal body is composed of a large number of interwoven hyphae. This mass of hyphae is called a **mycelium**.

The main body of the fungus can be found underground or embedded within the tissues upon which it is feeding. Multicellular fruiting bodies develop at specific times in the life cycle of most fungi.

Fungi are classified based primarily on their form of sexual reproduction. Many fungi go through asexual reproduction as well as sexual reproduction; however, some fungi primarily exhibit one or the other. In either case fungi reproduce through the production of spores. These are reproductive cells produced in a sporangium. There are many types of spores and sporangia in Kingdom Fungi. The names of the spores and sporangia vary based on the variations between different types of sporangia; however, spores usually have the same root name as the structure (sporangium) that produced them. Spores are resting stages that are resistant to a number of environmental fluctuations, including variations in temperature and external water.

Activities

■ Activity 21.1: Slime Molds

Slime molds are not true molds but are protozoans that alternate between a filamentous, mold-like growth pattern and an amoeboid growth form. Most of the life of a slime mold is spent in a state called a **plasmodium**, which is a large mass of cytoplasm with many nuclei. The plasmodium results from multiple rounds of mitosis without cytokinesis. It is the vegetative state of the organism and feeds by phagocytosis. This feeding pattern is not associated with true fungi.

When its environment begins to change, the plasmodium forms aggregates into masses of cytoplasm and prepares for sexual reproduction. At this time filaments are produced that extend vertically out of the masses and produce sporangia. Spores are produced within the sporangia and released. Meiosis follows the production of spores.

1. Observe the slime mold on display. Make note of its overall form and appearance.
2. Diagram the slime mold colony below.

COMPARE AND CONTRAST

1. Why do you believe slime molds are often included in discussions of fungi if they are not true fungi?

....

....

....

....

....

2. Was the colony you observed in a plasmodium state? How can you tell?

....

....

....

....

....

■ Activity 21.2: Zygomycetes

Zygomycetes are one of the more common groups of fungi. You have seen members of this group growing on stale bread. *Rhizopus stolonifer* is the most common of the bread molds. Members of this group have a simple body structure consisting of branched, coenocytic hyphae. The mycelium grows within and on the surface of whatever they are feeding upon. The mycelia grow and reproduce asexually under most conditions and are composed of haploid cells. The hyphae grow mostly horizontal; however, some grow down into the substrate on which the fungus is growing. Asexual reproduction is accomplished through **sporangia** that are produced on the tips of hyphae growing vertically away from the substrate. These are **sporangiophores**.

Sexual reproduction occurs only when the hyphae of two different mating types converge. The hyphae that fuse together form a gametangium at the point of contact. Within the gametangium, the cytoplasms of the two cells merge (plasmogamy). The nuclei from each cell fuse (karyogamy) and form a diploid zygote. The zygote is retained within the cell walls of the gametangium and develops into a thick-walled **zygospore** or, more technically, a **zygosporangium**. The cells within the zygospore divide by meiosis and haploid cells are released. No multicellular fruiting bodies are produced; sexual reproduction happens within the terminal cells of the fused hyphae.

1. Obtain a prepared slide of *R. stolonifer* in asexual stages. Observe the specimen under low power.
2. Locate a sporangium. Notice each sporangium has small, spherical spores within it. High power may be needed to observe details within the sporangium.
3. Diagram a sample of *Rhizopus*. Label the sporangium, sporangiophore, and hyphae.

4. Obtain a prepared slide of *R. stolonifer* in sexual stages. Observe the specimen under high power.

5. Locate a zygospore. It is supported on either side by swollen hyphae called **suspensors**. If you see earlier stages of sexual reproduction, the suspensors will be smaller and pro-gametangia form at the ends of the hyphae as soon as they contact one another.

6. Label a suspensor, hypha, and zygospore on **Figure 21-1**.

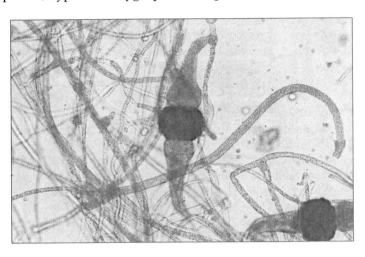

FIGURE 21-1 *Rhizopus stolonifer* showing sexual reproduction (×600). (Photo by James Richardson)

COMPARE AND CONTRAST

1. How are the zygospore and the sporangium similar? How are they different?

....

....

....

....

....

2. How does a hypha compare with a sporangiophore?

....

....

....

....

....

3. Which type of reproduction, sexual or asexual, in *Rhizopus* is most like what is seen in slime molds? Why?

....

....

....

....

....

■ Activity 21.3: Ascomycetes

Ascomycetes are common in many areas. You know them by their common names, such as truffles, yeasts, and powdery mildew. The ascomycetes are commonly referred to as the sac fungi due to their reproductive structures. They form spores within sporangia called **asci**. Each ascus is a sac-like cell in which haploid nuclei fuse to form a diploid zygote. That zygote immediately goes through meiosis, while inside the ascus, to produce haploid spores. These spores are called **ascospores**.

The fruiting body of an ascomycete is called an **ascocarp**. Ascocarps are found in three basic forms: an open, cup-shaped structure (apothecium); a closed, spherical body (cleistothecium); and a flask-shaped body with a small opening at one end (perithecium). Within the ascocarp are numerous asci, each producing spores. The ascocarp consists of a region of asci tightly packed together, called the **hymenium**, supported by the vegetative cells of the ascocarp.

1. Obtain a prepared slide of *Pezziza*. Observe the specimen under low power.

2. Find the ascocarp. What kind of ascocarp is this?

....

....

3. Sketch the general shape of the ascocarp below. Label the ascocarp and hymenium.

4. **Figure 21-2** shows an enlarged image of the hymenium. Label an ascus and a spore.

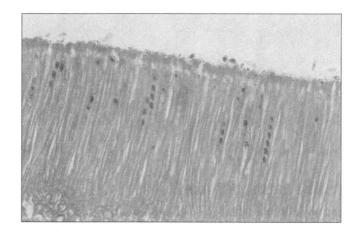

FIGURE 21-2   The hymenium of *Pezziza* (×100). (Courtesy of James D. Mauseth)

Yeasts are unusual ascomycetes because they are unicellular. They are best known for their use in producing alcoholic beverages and bread products. These uses are due to their metabolism. Yeasts are capable of fermentation when oxygen is not available for respiration. The byproducts of fermentation in yeasts are ethanol and carbon dioxide.

Because yeasts are unicellular they do not exhibit the same types of reproduction seen in other ascomycetes. They exhibit only asexual reproduction through **budding**. Budding is a type of cell division in which a smaller daughter cell, the **bud**, forms attached to the larger parent cells. Once the daughter cell releases from the parent cell, it enlarges to its mature size. At some later point it too will reproduce through budding.

5. Obtain a prepared slide of yeast cells (*Saccharomyces cerevisiae*). Observe the specimens under low power. Center a few in the middle of the field of view. Switch to high power.

6. Locate some budding cells. Diagram them below. Label the parent cell and the bud.

COMPARE AND CONTRAST

1. Asci produce sexual spores. What is the analogous structure in zygomycetes.

2. Compare an ascocarp with an ascus.

◼ Activity 21.4: Basidiomycetes

Basidiomycetes include many common fungi including mushroom and bracket fungi. They are commonly called club fungi because the reproductive structures that produce sexual spores are club-shaped. These structures are called basidia. Each basidium produces basidiospores that initially attach to the **basidium** through small, thin stalks called sterigmata (singular = sterigma). These **basidiospores** are subsequently released and carried by the wind.

The bulk of the mycelium grows underground. Spores germinate to produce haploid hyphae that form the primary mycelium. When hyphae from two mycelia come together, they go through plasmogamy to produce a **dikaryotic** cell with two haploid nuclei. This dikaryotic cell divides by mitosis and grows into a **secondary mycelium**. This secondary mycelium produces the fruiting body called the **basidiocarp**, which is better known as a mushroom. Dikaryotic hyphae extend upward from the soil initially in a small bud. The bud gradually enlarges as the mycelium grows and the mushroom takes shape.

Mushrooms are typical of the basidiocarps formed by this group of fungi, although some species produce other types of basidiocarps. The typical mushroom has several basic parts. Its stalk of a mushroom is called a **stipe**. At its apical end is an enlarged cap, or **pileus**. Initially, the pileus is attached to the stipe by an **inner veil**, a membranous structure that splits open when the pileus pulls away from the stipe to release the spores within. When that happens a ring of tissue called the **annulus** forms around the stipe from the remnants of the inner veil. Once the pileus has pulled away from the stipe, the gills, or **lamellae**, are visible on the lower surface of the pileus. The lamellae have numerous basidia attached to them and are the sites of spore production.

1. Obtain a cultivated mushroom. Identify as many of the structures above as possible on your specimen.
2. Diagram a cultivated mushroom below. Label the stipe, pileus, lamellae, and annulus.

3. Obtain a prepared slide of the cross-section of a mushroom (*Coprinus* sp.). Observe the specimen on scanning power.

4. Diagram the specimen below. Label the stipe, pileus, and lamellae.

5. Center one lamella in the field of view. Switch to high power. Observe the basidia.

6. Label the basidia, sterigma, basidiospore, and lamella on **Figure 21-3**.

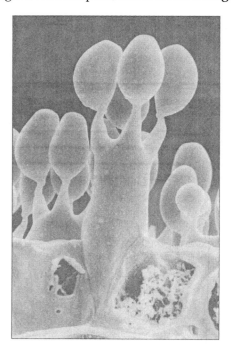

FIGURE 21-3   A representative basidium of basidiomycetes. (© Biophoto Associates/Photo Researchers, Inc.)

COMPARE AND CONTRAST

1. How are ascocarps and basidiocarps similar? How do they differ?

2. Why do you believe the stipe is a necessary part of a mushroom? Hint: Think about how the spores are dispersed.

....

....

....

....

....

▪ Activity 21.5: Imperfect Fungi

Imperfect fungi are those for which sexual reproduction has not been observed. They are classified within other divisions; however, we examine them separately to provide a closer examination of their features. Two of the most common and well-known imperfect fungi are *Aspergillus* and *Penicillium*. *Aspergillus* is used commercially to produce soy sauce. It is a common environmental fungus that can occasionally be pathogenic to humans. *Penicillium* is used to make many commercial products. The best known of these products is probably bleu cheese. The antibiotic penicillin was originally purified from *Penicillium*.

Both genera reproduce asexually through the production of **conidia**, asexual spores that are produced on the apical ends of **conidiophores**, a specialized type of aerial hyphae. The conidiophores of *Aspergillus* are rounded with short strings of spherical conidia extending off the outer surface. They look somewhat like a person's head whose hair is standing on end. In contrast, the conidiophores of *Penicillium* are fan-shaped filaments of conidia that attach to the apical end of the conidiophore. Some appear like miniature brooms and can be reminiscent of the bones in a human hand.

1. Obtain a prepared slide of *Aspergillus*. Observe the specimen under high power.
2. Locate a conidiophore. Sketch it below and label the conidiophore and conidia.

3. Obtain a prepared slide of *Penicillium*. Observe the specimen under high power.
4. Locate a conidiophore. Sketch it below and label the conidiophore and conidia.

COMPARE AND CONTRAST

1. Compare the structure of the conidiophores produced by *Aspergillus* and *Penicillium*.

....

....

....

....

....

2. How do these conidiophores differ from the sporangiophores of *Rhizopus*?

....

....

....

....

....

■ Activity 21.6: Lichens

Lichens are symbiotic relationships between fungi and algae, primarily green algae or cyanobacteria. There are three forms of lichens. **Crustose** lichens grow with the majority of the lichen embedded within the substrate, leaving only a thin section visible. **Foliose** lichens are flattened, overlapping membranous arrangements. They are sometimes referred to as a leaf-like pattern. **Fruticose** are the most complex form of lichens. These appear as miniature shrub-like growths.

   Internally, the structure of all lichens is quite similar. The upper surface is composed of a layer of fungal cells called the **upper cortex**. It is relatively thin. Just below the upper cortex is an **algal layer** housing the photosynthetic partner. The algal cells are closely associated with the fungal hyphae. Beneath the algal layer is a thick section of fungal hyphae called the **medulla**. The lowest edge of the medulla borders the **lower cortex**, which serves to anchor the lichen to its substrate.

   In nature lichens typically reproduce asexually through fragmentation. The fungus reproduces sexually; however, the spores must come into contact with an appropriate algal symbiont or the fungus will not survive.

1. Observe the lichens on display. Identify crustose, foliose, and fruticose specimens. Diagram or describe them below.

2. Obtain a prepared slide of the cross-section of a lichen. Diagram it below and label the upper cortex, lower cortex, medulla, and algal layer.

COMPARE AND CONTRAST

1. Compare the location of the algal layer with the function of the algae.

2. In what ways are lichens similar to fungi in external features? In what ways are they different?

■ Activity 21.7: Mycorrhizae

Plant roots form mutualistic relationships with fungi called mycorrhizae. The fungus increases water and phosphorus absorption that the plant acquires from the fungus. The fungus in turn acquires nutrients from the plant's phloem.

■ Activity 21.8: Rusts and Smuts

Rusts and smuts are fungal infections of plants. We examine wheat rust as our representative species. Wheat rust (*Puccinia graminis*) is a parasite of grains, including wheat, oats, and barley. It has a complex life cycle (**Figure 21-4**) involving two different host plants and three different types of spores.

The **basidiospores** of wheat rust are released in the spring. These spores do not survive long unless they come into contact with the young leaves of a barberry plant. If the leaf

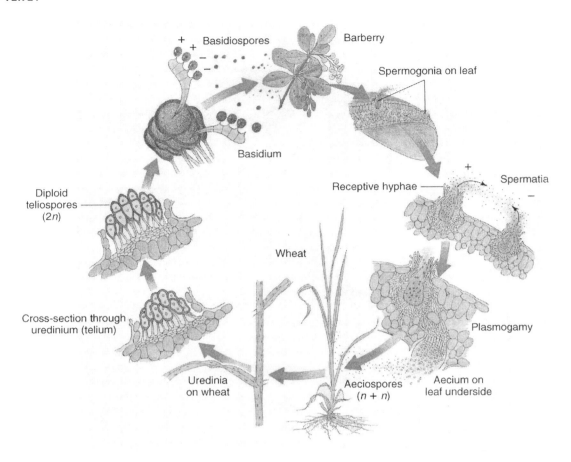

FIGURE 21-4 Life cycle of wheat rust (*Puccinia graminis*). (© Jones & Bartlett, LLC)

surface is moist, the basidiospores germinate and produce hyphae that penetrate into the leaf's tissues. As the mycelium continues to develop, at some point it produces spermogonia, masses of hyphae on the surface of the barberry leaf. The spermogonia produce small cells called **spermatia** that must come into contact with a receptive hypha, also embedded within **spermogonia**, of another mating type. This is similar to the requirements for sexual reproduction in zygomycetes.

Aeciospores are formed on the lower surface of the barberry leaf in an inverted cup-shaped structure called an **aecium**. Around the same time as aeciospore formation, the spermogonia form on the upper surface of the same leaf.

1. Obtain a prepared slide of the three stages of wheat rust infection. Observe the specimen under high power. Locate the stage with spore production on the lower surface of the barberry leaf.

2. Diagram an aecium with aeciospores and label them below.

The dikaryotic aeciospores are released and carried by air currents. Those that land on wheat leaves and enter an open stoma survive and germinate. These grow quickly within the wheat and produce a **uredinium** underneath the upper epidermis of the leaf. The uredinium produces another type of spore, the dikaryotic **urediniospores**. These spores are produced rapidly and result in the epidermis of the leaf becoming separated from the mesophyll. The opening of the epidermis allows the urediniospores to be released. The urediniospores are produced to rapidly infect the leaves of other wheat plants or other leaves of the same plant.

3. Observe the uredinium and urediniospores on the wheat rust slide.

4. Diagram the uredinium below. Label the urediniospores, uredinium, and leaf epidermis.

Toward the end of wheat's growing season, a third type of spore, called **teliospores**, are produced on the surface of wheat leaves in structures called **telia**. Telia form from uredinia that stop producing urediniospores. Teliospores survive the winter in straw or soil. In the spring the dikaryotic nuclei in each teliospore fuse to form a diploid cell. That cell immediately divides by meiosis to produce the four haploid cells. These cells then germinate in the spring to produce basidia. The basidiospores that are subsequently produced will in turn infect a barberry leaf.

5. Observe a telium and teliospores on the surface of a wheat leaf. Teliospores are produced at the tips of vertical structures extending out of the surface of the wheat leaf. They are larger and have thicker walls than the urediniospores.

6. Diagram a telium with teliospores below and label each.

7. If available, observe some preserved specimens of rusts and smuts.

Observe the preserved fungal specimens on display and be able to match them with their appropriate groups. Your instructor will provide additional instructions as necessary.

Study Guide

- Be able to define the terms in bold.
- Understand the role of spores in fungal reproductive strategies.
- Be able to differentiate between the various types of fungal spores.
- Be able to describe the roles of fungi and algae in a lichen.
- Be able to identify the three forms of lichens.

Conclusions

1. Define the following terms.
   a. Basidiospore

   b. Conidia

   c. Zygospore

2. What is a dikaryotic cell?

3. Describe the stipe, pileus, and lamellae of a mushroom. Which is directly involved in reproduction?

4. Compare and contrast hyphae and mycelia.

5. Describe the conditions necessary for sexual reproduction in *Rhizopus*. In which fungal group is *Rhizopus* placed?

6. Why are *Penicillium*, *Aspergillus*, and other members of the imperfect fungi considered to be "imperfect?"

7. Compare and contrast asexual reproduction in yeast with that of *Penicillium*.

8. List three characteristics shared by all fungi.

9. Why are slime molds not considered to be true fungi? Why are they included here?

10. What are the hosts needed for *Puccinia graminis* reproduction?

Chapter 22

© Cristi Matei/ShutterStock, Inc.

Prokaryotes

Laboratory Activities

Activity 22.1: Culturing Bacteria

Activity 22.2: Staining Bacteria

Activity 22.3: Pour Plate Method of Bacterial Enumeration

Goals

Following this lab students should be able to

- Identify the cellular characteristics of bacteria.
- Differentiate between prokaryotes and eukaryotes.
- Understand the importance of the Gram stain in identifying bacterial isolates.
- Describe the environmental prevalence of prokaryotes.

Introduction

Prokaryotes are unicellular organisms that lack nuclei in their cells. At a cellular level there are many differences between prokaryotes and eukaryotes. The lack of a nucleus is one of the most distinctive of these; however, prokaryotic cells also lack the internal membranes that form the endoplasmic reticulum, Golgi apparatus, and many other organelles common to eukaryotes. Although these organisms have relatively simple cells, they are numerous and quite successful in many environments.

Two domains of organisms are prokaryotic. **Domain Archaea** includes numerous organisms that are sometimes referred to as extremophiles. Archaeans live in environments that exhibit extreme characteristics, such as salt flats, deep ocean vent, sulfuric hot springs, and anaerobic sediments. None of the archaeans is pathogenic (disease-causing); however, they are important because of the role they serve in their respective ecosystems.

Domain Bacteria contains the more familiar prokaryotes. Some of these are **pathogenic** and may cause a variety of diseases in humans and other organisms. Many bacteria are beneficial to humans, either as decomposers in the environment, part of the nitrogen cycle, or as mutualistic organisms living in or on the human body. It is important to remember that not all bacteria are harmful. In fact, the human body relies heavily on some bacteria and does not function well without them, for example, the bacteria living in the human intestinal tract produce vitamins our bodies require. Many species are present in the large intestine, for instance, the familiar *Escherichia coli*. The majority of *E. coli* strains are nonpathogenic, and many are beneficial to the human body by producing vitamin K, which our bodies readily absorb. You are probably more familiar with the minority of *E. coli* strains that are pathogenic. Some strains can cause food poisoning and have caused some high-profile outbreaks.

Bacteria are incredibly numerous and diverse in our world. They often go unnoticed due to their microscopic size, but there are millions present in a single gram of soil. They can live in aquatic environments, both fresh water and marine; in terrestrial environments; and within the tissues of other organisms. Some are predatory and feed on other prokaryotes. Some are autotrophic and produce carbohydrates through photosynthesis or chemosynthesis (using chemical energy to power carbon dioxide reduction). A few of these chemosynthetic species live in the soil and can transform gaseous nitrogen in the atmosphere into forms more usable by other organisms, such as ammonia, nitrites, or nitrates. Those that are photosynthetic can produce oxygen as a byproduct in a process similar to plant photosynthesis, whereas some use pigments other than chlorophyll and may or may not produce oxygen. Some are parasitic and live off the tissues of other organisms. Some are saprotrophic and feed on dead and decaying organic matter. These saprotrophs are important in the chemical cycling that occurs in all ecosystems.

Bacteria have many characteristics that can be used to classify them into different families or species. Some of these are biochemical in nature, such as the ability to produce enzymes to break down gelatin or ferment lactose, and some are structural, such as the type of cell wall present. The vast majority of bacteria have cell walls, and the structural component of all bacterial cell walls is a polysaccharide called **peptidoglycan**. Bacteria can be found in numerous shapes; however, the three most common are spherical cells called **cocci** (singular = coccus), rod-shaped cells called **bacilli** (singular = bacillus), and spiral-shaped cells called **spirilla** (singular = spirillum). Some species of bacteria are found as individual cells. Others form species-specific arrangements. For example, cells may occur in pairs (**diplo-**), chains (**strepto-**), or irregular clusters (**staphylo-**). These terms are used in conjunction with the shape of the cells to define the morphology and arrangement of a bacterial sample. *E. coli*, for example, is a diplobacillus.

Activities

■ Activity 22.1: Culturing Bacteria

Bacteria can be cultured and grown in laboratory settings using various types of media. Liquid culture media, referred to as broths, are used to grow large quantities of bacteria in a single container. They are convenient for use on smaller scales when a pure culture of bacteria is being grown. A **pure culture** contains only one species or subspecies of bacteria. If you are trying to grow bacteria from a mixed source, such as an environmental sample, then you will typically use a solid medium. Solid media contain a compound called **agar**. Agar liquefies at 100°C and solidifies at 45°C. Few organisms can break down agar. It is, therefore, a convenient additive to make a culture medium solid for use in Petri dishes.

A solid media that is **nonselective** supports the growth of most bacteria. It is a complex medium containing a variety of carbon sources. When bacteria are grown on solid media, they typically form colonies. A colony is a mass of bacteria that have originated from one or a few bacteria that landed in that location on the agar. As they continue to divide, the cells form a larger, visible mass. Colonies are usually round in shape, but some can be irregular or take on other shapes.

1. Obtain two tryptic soy agar plates, one test tube containing sterile water, and four sterile cotton-tip applicators.

2. Using a permanent marker or wax pencil, draw a line down the center of the Petri dish dividing the plate into halves. Always label your plate on the bottom (agar side) and not the lid.

3. Label one plate "body" and the other "environmental."

4. Take two samples from your person for the body plate. Take the first sample from your skin. Dampen the cotton-tip applicator in the sterile water. Remove the excess water by rolling the cotton end against the side of the test tube before removing it from the sterile water tube. Recap the sterile water. Gently swab an area of your skin (elbow, behind your ears, along the side of your nose, etc.). Then swab the cotton-tip applicator across half of the plate. Roll the applicator to ensure that all sides of the cotton tip come into contact with the agar. Do not press too hard against the agar. It is soft and easy to puncture. Discard the used cotton-tip applicator as directed. Label that half of the plate with the area swabbed.

5. Take the next sample from inside your mouth. It is not necessary to dampen the cotton tip for this exercise. Simply remove the cotton-tip applicator from its package and swab the inside of your mouth (cheeks or along the gum line). Then swab the cotton tip along the surface of the agar on the other half of the plate. Label that half of the plate with the source sampled. Discard the cotton-tip applicator as directed.

6. For the environmental plate you will need two sterile cotton-tip applicators and one vial of sterile water. Again, dampen each cotton-tip applicator as above, use the dampened applicator to sample an area, and then swab the applicator across the surface of the agar. You will sample two environmental locations. Be sure to label each half of the plate with the source sampled.

7. The plates must be incubated at 25 to 37°C for 24 to 48 hours. Always incubate the plates in an inverted position (bottoms up!).

8. Record the bacterial growth results in **Table 22-1**.

TABLE 22-1: Bacterial Samples			
	Colony Color	Shape	Number Present
Body Samples			
1.			
2.			
3.			
4.			
5.			
6.			
Environmental Samples			
1.			
2.			
3.			
4.			
5.			
6.			

9. Based on these results, can you determine which species of bacteria are present? Why or why not?

....

....

....

COMPARE AND CONTRAST

1. Do you see the same number and types of organisms on both plates (environmental and body)?

....

....

....

....

2. Describe some of the differences in how each colony appears.

....

....

....

....

3. Did you observe more bacteria from the body samples or from the environmental samples? What factors may have contributed to this?

....

....

....

....

■ Activity 22.2: Staining Bacteria

Two main groups of bacteria can be identified based on the type of cell wall present. The difference is determined using a stain called the **Gram stain**. **Gram-positive** bacteria have a thick layer of peptidoglycan cell wall surrounding their plasma membranes. **Gram-negative** bacteria have a thin layer of peptidoglycan cell wall surrounding the plasma membrane and have an additional outer membrane surrounding the peptidoglycan layer. Both groups of bacteria contain some species that are beneficial and some that are pathogenic. Both types of cell walls have the same major functions.

They help prevent osmotic lysis in the same way that a plant cell wall prevents excessive amounts of water from entering the plant cell. Both types of cell walls help maintain the shape of the bacterium, can provide binding sites for attaching to substrates, and may contain structures that help protect the bacterium from the mammalian immune response. In spite of these common functions, the structural differences do give each type of bacterial cell some properties not shared by the other group. For example, the antibiotic penicillin damages the cell wall of a gram-positive bacterium but does not harm a gram-negative bacterium. Likewise, tetracyclines are active against gram-negative bacteria but not against gram-positive bacteria.

The Gram stain uses **crystal violet** as the primary stain. It embeds into the peptidoglycan layer. **Iodine** is added as a **mordant**, a chemical that interacts with the dye. In this case the iodine–crystal violet combination forms crystals that get stuck in the peptidoglycan layer of a gram-positive cell. **Ethanol** is used to wash any unbound stain out of the cell wall. This removes the crystal violet from a gram-negative cell wall. Because bacteria are so small, they are difficult to see without a stain. To counter this problem we use a secondary stain, in this case **safranin**, to make the gram-negative cells visible.

1. Obtain an agar plate containing *Escherichia coli* or *Staphylococcus aureus*, a slide, and an inoculating loop.
   a. If the loop is not sterile, you will need a Bunsen burner or equivalent to sterilize it.
   b. Hold the wire loop in the flame of the burner until the wire glows orange-red. Allow it to cool for 20 to 30 seconds before use.
2. Draw a circle on the bottom of the slide approximately 1 inch in diameter.
3. Place a small drop of water on the top of the slide in the middle of the circle.
4. Touch the tip of the sterile loop to the bacterial culture growing on the surface of the agar. Then swirl the loop into the water and spread the water out on the slide, staying inside the lines of the circle if possible. The circle is there to help you find the sample under the microscope.
5. Allow the slide to air dry.
6. Heat fix the slide by passing the slide through the open flame of a Bunsen burner three times. This prevents the bacteria from washing off the slide and ensures all the bacteria are dead.
7. Cover the slide with crystal violet for 30 seconds. Rinse with water.
8. Cover the slide with Gram's iodine for 10 seconds. Rinse with water.
9. Destain with ethanol for 2 to 3 seconds. Rinse with water. This is a sensitive step. Too little or too much exposure to ethanol can be problematic.
10. Cover the slide with safranin for 30 seconds. Rinse with water. Blot dry with bibulous paper.
11. Observe your slide under the microscope. Begin with the scanning objective and work your way up to the oil immersion (100×) lens. You need to use microscope oil for this. Your instructor will provide you with additional instructions.
12. What color are the bacteria that you see?
13. Gram-positive bacteria stain purple, whereas gram-negative bacteria appear pink. Is your sample gram-positive or gram-negative?

COMPARE AND CONTRAST

1. Describe the difference between the data from culturing organisms and from staining.
2. Which is most useful for identifying a particular organism, culture on tryptic soy agar or staining? Why?
3. Would either of these techniques give you enough information to identify an organism for sure? Why or why not?

■ Activity 22.3: Pour Plate Method of Bacterial Enumeration

Bacteria are plentiful in virtually any environment on earth. We are able to culture many bacteria from soil, water, and other environments; however, many of the bacteria present have never been grown in a laboratory. Those can only be detected using genetic tests. You will be analyzing a soil sample to approximate the number of bacteria present. Remember, this will be a conservative estimate because many more species of bacteria do not grow in culture.

1. Obtain a soil sample, two bottles with 99 ml of sterile water, and four sterile Petri dishes.
2. Measure 1 gram of a soil sample. Add this to one bottle of water. Mix. (This is a 1:100 dilution).
3. Make another 1:100 dilution by adding 1 ml of the mixture from the first bottle to the second bottle. This is a total dilution of 1:10,000 (1:100 × 1:100 = 1:10,000).
4. Label the plates 1:100, 1:1,000, 1:10,000, 1:100,000
5. Pipette 1 ml of the 1:100 dilution into the Petri dish labeled 1:100.
6. Pour 19 ml of melted nutrient agar into the Petri dish.
7. Swirl the plate to mix the solution and media. Set the plate aside to cool.
8. Pipette 0.1 ml of the 1:100 dilution into the plate labeled 1:1,000. Repeat steps 6 and 7.
9. Pipette 1 ml of the 1:1,000 dilution into the plate labeled 1:10,000. Repeat steps 6 and 7.
10. Pipette 0.1 ml of the 1:10,000 dilution into the plate labeled 1:100,000. Repeat steps 6 and 7.
11. Incubate the plates inverted (bottoms up!) for 24 to 48 hours at 35 to 37°C.

Count the number of colonies during the following lab period. Record your results in **Table 22-2**. You will also need to calculate the **colony-forming units** per gram (CFU/g). We use the term CFU to reflect the number of bacteria present in the original sample because each colony could have arisen from as little as one bacterium; however, multiple bacteria could have landed in the same location and contributed to the same colony. To calculate CFU/g, use the following formula: CFU/g = number of colonies/(dilution factor of the bottle × volume plated). Note: the dilution factor × volume plated is the total dilution (labeled on each plate).

Statistically relevant colony counts are between 25 and 250 colonies per plate. If you count fewer than 25 colonies, record your results as too few to count (TFTC). If you count more than 250 colonies, record your results as too numerous to count (TNTC).

TABLE 22-2: Pour Plate Results		
Plate	Number of Colonies	CFU/g
1:100		
1:1,000		
1:10,000		
1:100,000		

COMPARE AND CONTRAST

1. How does the type of information gained from the pour plate differ from that obtained by swabbing samples onto nutrient agar?

2. Does the information gained from the pour plate method bring you any closer to identifying a species of bacteria than the previous two methods? Why or why not?

3. Of the four plates, which showed the most desired number of colonies? What does that tell you about the benefits of diluting the samples?

Study Guide

- Be able to define the terms in bold.
- Be able to describe the roles of prokaryotes.
- Be able to compare and contrast the usefulness of the three methods presented.
- Be able to answer each question in the lab exercise.

Conclusions

1. What two groups of organisms are prokaryotic?

2. List two ways in which prokaryotes differ eukaryotes.

3. Describe one advantage of each method used today.

4. What is one limitation of growing bacteria on the nutrient agar?

5. Describe two reasons why staining is useful in studying bacteria?

6. Describe the advantage of the pour plate relative to the other two methods described.

7. Describe at least one similarity between plant cells and bacteria.

8. Where in a plant might you find bacteria?

CPSIA information can be obtained at www.ICGtesting.com
Printed in the USA
BVOW04s1018240116

434042BV00034B/1236/P